WOMEN'S ISSUES

FEDERAL GRANT-MAKING TO WOMEN IN STEM RESEARCH FIELDS

ANALYSIS AND REPRESENTATION PROPOSALS

WOMEN'S ISSUES

Additional books in this series can be found on Nova's website under the Series tab.

Additional e-books in this series can be found on Nova's website under the eBooks tab.

WOMEN'S ISSUES

FEDERAL GRANT-MAKING TO WOMEN IN STEM RESEARCH FIELDS

ANALYSIS AND REPRESENTATION PROPOSALS

ROCHELLE SCHULER
EDITOR

NOVINKA

New York

NOTICE TO THE READER

Library of Congress Cataloging-in-Publication Data

ISBN: 978-1-63485-679-9

Published by Nova Science Publishers, Inc. † New York

CONTENTS

PREFACE

In fiscal year 2014, U.S. universities received nearly $25 billion in federal grant funding for Science, Technology, Engineering, and Mathematics (STEM) research. Studies show women are largely underrepresented in STEM fields. Federal agencies are required to enforce Title IX—a law prohibiting discrimination on the basis of sex in education programs receiving any federal financial assistance—including at universities they fund. GAO was asked to provide information on federal grant-making to women in STEM. This book examines the extent to which differences exist in federal grant awards between women and men in STEM fields; the extent to which federal agencies enforce Title IX at universities they fund for STEM research; and possible actions federal agencies could take to address the representation of women in STEM research.

In: Federal Grant-Making to Women … ISBN: 978-1-63485-679-9
Editor: Rochelle Schuler

Chapter 1

WOMEN IN STEM RESEARCH: BETTER DATA AND INFORMATION SHARING COULD IMPROVE OVERSIGHT OF FEDERAL GRANT-MAKING AND TITLE IX COMPLIANCE*

United States Government Accountability Office

WHY GAO DID THIS STUDY

In fiscal year 2014, U.S. universities received nearly $25 billion in federal grant funding for Science, Technology, Engineering, and Mathematics (STEM) research. Studies show women are largely underrepresented in STEM fields. Federal agencies are required to enforce Title IX—a law prohibiting discrimination on the basis of sex in education programs receiving any federal financial assistance—including at universities they fund. GAO was asked to provide information on federal grant-making to women in STEM.

This report examines: (1) the extent to which differences exist in federal grant awards between women and men in STEM fields, (2) the extent to which federal agencies enforce Title IX at universities they fund for STEM research, and (3) possible actions federal agencies could take to address the

* This is an edited, reformatted and augmented version of a United States Government Accountability Office, Publication No. GAO-16-14, dated December 2015.

representation of women in STEM research. GAO analyzed data on all STEM research grants made in fiscal years 2009 through 2013—the most recent data available—by the six federal agencies that provided 90 percent of STEM research funding in fiscal year 2012 through 2014; reviewed literature, federal laws and regulations, and agency documents; interviewed federal officials; and consulted 19 STEM diversity experts.

WHAT GAO RECOMMENDS

GAO recommends that DOD, DOE, and NASA collect additional data; DOD and HHS conduct Title IX compliance reviews; and DOJ facilitate information sharing among STEM agencies. Agencies agreed in principle, but some cited potential implementation challenges. GAO maintains action is feasible and warranted as discussed in the report.

WHAT GAO FOUND

GAO's analysis of Science, Technology, Engineering, and Mathematics (STEM) research grant awards made between fiscal years 2009 and 2013 identified no disparities in success rates between women and men at three agencies selected for review, but data limitations provided limited insight into success rates at three other agencies selected for review. At two of the agencies with data limitations— the Departments of Defense (DOD) and Energy (DOE)—GAO found evidence of disparities in success rates for women and men within certain agency components. Data limitations at the National Aeronautics and Space Administration (NASA) prevented GAO's analysis of success rates altogether. This lack of complete, linked electronic proposal and award data at NASA and some components at DOD and DOE impacts their ability to fully evaluate their programs' performance against their stated goals of funding the most qualified scientists, irrespective of gender. Adhering to federal internal control standards regarding data collection of an entire process can ensure that these agencies have the data needed for effective program management and monitoring.

Source: GAO analysis of STEM grant data for fiscal years 2009 through 2013 from Health and Human Services (HHS) National Institutes of Health (NIH); National Science Foundation (NSF); Department of Agriculture (USDA) National Institute of Food and Agriculture (NIFA); Department of Defense (DOD); Department of Energy (DOE); and NASA. | GAO-16-14.

Success Rates between Women and Men at Federal STEM Grant-making Agencies.

Two of six agencies GAO reviewed that fund STEM research at universities— DOD and the Department of Health and Human Services (HHS)—are not conducting required Title IX compliance reviews. Since HHS oversees Title IX compliance of National Institutes of Health (NIH) funding recipients, which account for the bulk of STEM research grantees, billions of federal research dollars may not be subject to potential Title IX oversight. The Department of Justice (DOJ) is designated by Executive Order to coordinate Title IX compliance across federal agencies, including information sharing, but it has no formal information sharing process among STEM agencies. Officials at five of the six agencies GAO interviewed reported a desire for DOJ to facilitate interagency information sharing on Title IX best practices for compliance activities. Without such information sharing, these STEM agencies may miss opportunities to improve their compliance programs and coordinate with each other.

GAO identified through a literature review and expert interviews 13 potential actions federal agencies could take to address the underrepresentation of women in STEM research. These actions fell into four areas: (1) enhancing agency leadership and collaboration, (2) establishing family-friendly policies for grantees, (3) overseeing the research proposal review process, and (4) funding and assisting academic institutions. While not all of the actions GAO identified are relevant to or feasible for each agency, all six agencies in GAO's review indicated that they are either taking actions in some of these areas, or would be willing to explore their applicability. Most of the agency officials GAO spoke with acknowledged the potential benefits of these actions.

ABBREVIATIONS

DOD	Department of Defense
DOE	Department of Energy
DOJ	Department of Justice
Education	Department of Education
HHS	Department of Health and Human Services
NASA	National Aeronautics and Space Administration
NIFA	National Institute of Food and Agriculture
NIH	National Institutes of Health
NSF	National Science Foundation
OMB	Office of Management and Budget
SDR	Survey of Doctorate Recipients
SSA	Social Security Administration
STEM	Science, technology, engineering, and mathematics
Title IX	Patsy Takemoto Mink Equal Opportunity in Education Act
USDA	Department of Agriculture

* * *

December 3, 2015

The Honorable Eddie Bernice Johnson
Ranking Member
Committee on Science, Space, and Technology
House of Representatives

The Honorable Rosa DeLauro
House of Representatives

The Honorable Louise M. Slaughter
House of Representatives

Since the enactment of Title IX in 1972—which prohibits discrimination on the basis of sex in education programs and activities receiving any federal financial assistance[1]—women have made significant gains in many academic fields. Nevertheless, recent research shows that they continue to lag behind men in academic and professional advancement in the majority of science, technology, engineering, and mathematics (STEM) fields.[2] In fiscal year 2014, nearly $25 billion in federal funding for research in STEM fields was awarded to colleges and universities around the country.[3] Federal agencies providing funding to universities are responsible for enforcing Title IX compliance with respect to those universities, including conducting periodic compliance reviews of their funding recipients.[4]

You requested information regarding federal grant-making to women and men for research in STEM fields. In March 2015, we published a report on the extent to which federal agencies collect data that could be used to analyze any differences in federal grants to women and men in STEM fields.[5] This report is our final response to your request related to the representation of women in federal STEM research programs. Specifically, our objectives were to determine (1) the extent to which differences exist in federal research grant awards between women and men in STEM fields and what factors might explain any differences, (2) the extent to which federal agencies enforce Title IX at universities they fund for STEM research, and (3) possible actions federal agencies could take to address the representation of women in federal STEM research.

To address these objectives, we included the six federal STEM research grant-making agencies in our review that together funded 90 percent of the federal government's investment in basic and applied research in STEM fields from fiscal year 2012 through 2014: the Department of Agriculture-National Institute of Food and Agriculture[6] (USDA-NIFA), the Department of Defense (DOD), the Department of Energy (DOE), the National Aeronautics and Space Administration (NASA), the National Institutes of Health[7] (NIH), and the National Science Foundation (NSF). This report focuses on federal grant funding for research at universities in the core STEM fields of engineering, life sciences (e.g., agricultural, biological, and environmental sciences), physical sciences (e.g., chemistry, earth sciences, and physics), computer and information technology, and mathematics and statistics.[8] We focused on core STEM fields in this report because data show women continue to be underrepresented in the majority of these fields.[9] We collected and analyzed data and information through several methods. First, we reviewed relevant federal laws, regulations, and guidance. We then conducted a comprehensive literature review to identify factors that may contribute to women's representation in academic STEM fields, as well as possible actions federal agencies could take in this area beyond Title IX enforcement activities.

To determine if differences existed in federal grant-making to women and men in STEM fields, we obtained available administrative data from each agency included in our review for all research grant proposals received and awards made in fiscal years 2009 through 2013, the most recent complete year of agency data available when we began our review.[10] Through consultations with agency officials responsible for the data systems and by reviewing these data, we assessed the reliability of these data and included in our analysis only those data that we found to be sufficiently reliable for the purposes of our report. To complement our analysis of the agency administrative data and identify factors that may contribute to individuals' success obtaining federal funding, we separately analyzed data from the National Science Foundation's Survey of Doctorate Recipients (SDR) from 2013 for four of the six agencies.[11] Through interviews with NSF officials and a review of survey methodology documents, we also assessed the reliability of these data and found them to be sufficiently reliable for the purposes of this report. We evaluated each agency's data collection practices by assessing the extent to which the data collected would allow agencies to evaluate their programs' performance against their own stated program goals. We also compared their data collection practices against federal internal control standards, which outline a set of key elements of record keeping which agencies should have in place in order to ensure effective program management and also to allow for internal and external evaluation and oversight.[12]

To assess the extent to which the six federal agencies enforce compliance with Title IX at universities they fund for STEM research, we interviewed officials responsible for civil rights compliance at the grant-making agencies and reviewed agency policies and documents related to their Title IX compliance activities. We also interviewed officials at the Department of Education (Education) and the Department of Justice (DOJ) about their roles and responsibilities with regard to Title IX compliance. We evaluated the Title IX compliance activities of these six STEM grant-making agencies against relevant statutory and regulatory requirements in addition to federal internal control standards for effective communication.[13] Finally, we identified possible actions federal agencies could take to improve the representation of women in federal STEM research through a review of the literature and by consulting 19 subject matter experts on our list of possible actions.[14] We then interviewed officials at the grant-making agencies to determine the extent to which the agency takes any of these actions. For more information on our scope and methodology, please see appendix I.

We conducted this performance audit from October 2014 to December 2015 in accordance with generally accepted government auditing standards. Those standards require that we plan and perform the audit to obtain sufficient, appropriate evidence to provide a reasonable basis for our findings and conclusions based on our audit objectives. We believe that the evidence obtained provides a reasonable basis for our findings and conclusions based on our audit objectives.

Background

The six federal agencies included in our analysis provide billions of dollars annually for university research in STEM fields, with NIH providing nearly twice the amount of funding than the other five agencies we reviewed combined.[15] Figure 1 details the total amount of research funding provided to universities by each agency in fiscal year 2014.

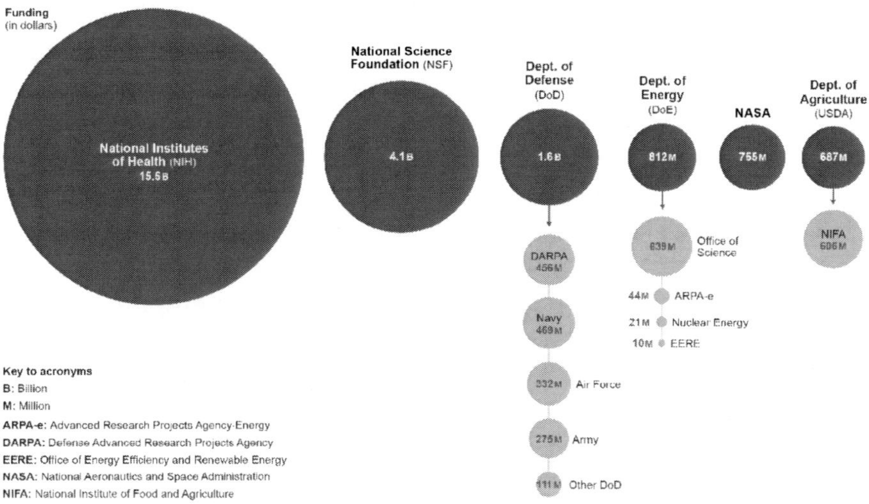

Source: National Science Foundation, National Center for Science and Engineering Statistics, Survey of Federal Funds for Research and Development, Fiscal Year 2014. | GAO-16-14.

Notes: As noted, these funding figures are drawn from the 2014 NSF Survey of Federal Funds for Research and Development and therefore may differ from agency figures reported elsewhere. FY 2014 figures include only research obligations in fields that fall within the scope of our review, i.e., computer and information technology, engineering, environmental sciences, life sciences, mathematics and statistics, physical sciences, and other sciences not elsewhere classified. These may differ from total agency research funding obligations for FY 2014. For example, NSF's total STEM research obligations include funding for additional STEM fields such as psychology and social sciences.

Figure 1. Fiscal Year 2014 Federal Obligations from Six Agencies for Select Areas of STEM Research Performed at Universities and Colleges.

Federal Grant Awards and Grant Life Cycle

While there can be significant variation among different grant programs, most federal grants share a common life cycle for administering the grants: pre-award, award, implementation, and closeout (see fig. 2). Among the agencies we reviewed, with some exceptions, during the pre-award stage, applicants may initiate contact with the program officer at the relevant funding agency before submitting a proposal to discuss their potential research ideas and obtain input on if their idea fits with the agency's funding goals.[16] The degree of formality of this pre-award stage communication between the program officer and applicant varies by agency and component; at some agencies and for some grant programs, it is required and at others it is optional. During the award stage, the federal awarding agency enters into an agreement with grantees stipulating the terms and conditions for the use of grant funds, including the period of time funds are available for the grantee's use. During the post-award stage, the grantee carries out the requirements of the agreement and requests payments, while the awarding agency approves payments and oversees the grantee. Once the grantee has completed all the work associated with a grant agreement or the end date for the grant has arrived or both, the awarding agency and grantee close out the grant.

Title IX of the Education Amendments of 1972

Title IX of the Education Amendments of 1972[17] is the primary federal law that addresses sex discrimination in all federally funded grant programs at educational institutions.[18] All federal agencies that provide funding for education programs and activities, which includes STEM funding to universities, have enforcement responsibilities under Title IX, including issuing regulations,[19] conducting periodic compliance reviews at these institutions,[20] and investigating timely written complaints of sex discrimination against these recipients.[21] Recipients of federal assistance—in this case, university grantees—also have some compliance responsibilities.[22] Figure 3 outlines the various compliance activities required under Title IX and the entity responsible for carrying out each activity.

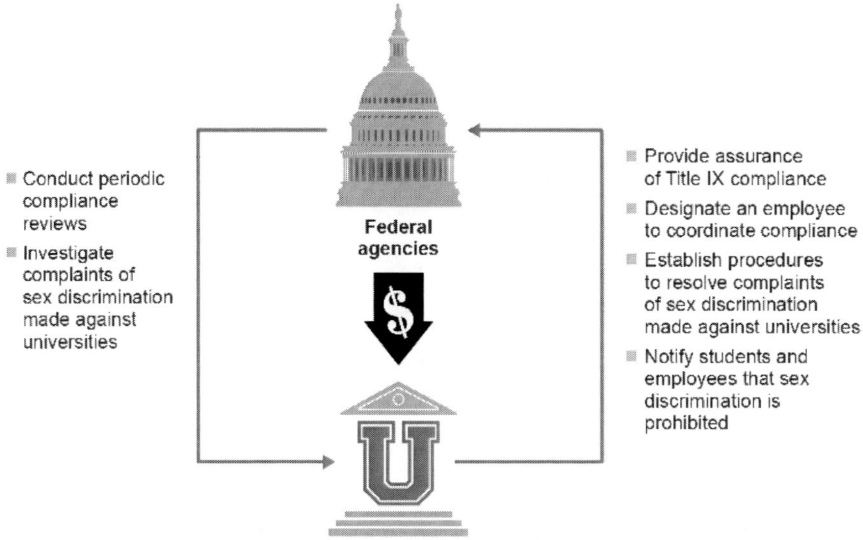

Source: GAO analysis of agency Title IX regulations. | GAO-16-14.

Figure 3. Key Title IX Requirements for Federal Funding Agencies and Universities.

A Title IX compliance review is an agency's assessment of whether a grantee is complying with the law. Agencies may initiate compliance reviews in the absence of a complaint or expand a complaint investigation into a compliance review of issues outside of the complaint. Such an assessment can be conducted on-site or via a desk audit. Federal agencies may also coordinate with each other to carry out their compliance responsibilities. Education also plays a key role in ensuring compliance with Title IX because it provides funding to most universities in the United States and is also the primary federal agency that receives, investigates, and resolves Title IX complaints. Finally, under Executive Order 12250, DOJ has responsibility for playing a leadership role in coordinating the "consistent and effective implementation" of several civil rights laws, including Title IX.[23]

What Is Known About Women's Participation in STEM Research

While data show that women's representation in STEM fields has risen since Title IX was enacted, women continue to lag behind men in most STEM fields in both educational and professional achievement. While women received more degrees overall than men in the 2011-2012 academic year, they received fewer degrees in all core STEM fields except the life sciences (see fig. 4).

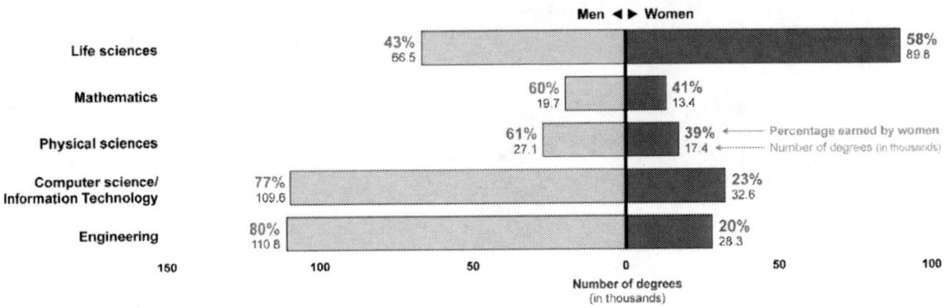

Source: GAO analysis of data from the Integrated Postsecondary Education Data System, GAO-14-374. | GAO-16-14.

Notes: In this figure, degrees include less than bachelor's, bachelor's, post-bachelor's certificates, master's, and doctorate/professional. Percentages for each discipline may not add to 100 percent due to rounding.

Figure 4. Degrees Awarded in Core Science, Technology, Engineering, and Mathematics (STEM) Fields, by Gender, 2011-2012 Academic Year.

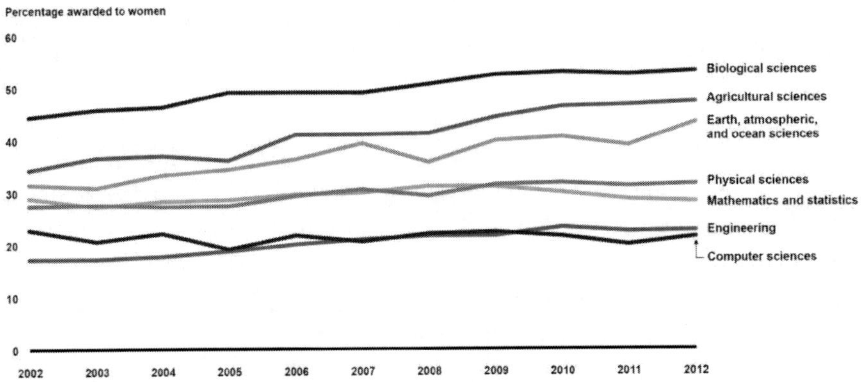

Source: GAO analysis of data from National Science Foundation, National Center for Science and Engineering Statistics. | GAO-16-14.

Notes: Data in this figure are the most recent data available.

Figure 5. Percent of Doctoral Degrees Awarded to Women, by Field, Fiscal Years 2002–2012.

Similarly, from 2002 through 2012, women received one-third or fewer of the doctorates awarded in physical sciences, mathematics, engineering, and computer science (see fig. 5).[24] Although the proportion of new PhDs in mathematics and computer science who were women declined slightly from 2002 to 2012, the total number of women earning doctorates in these two fields increased slightly during the same time period.

Lastly, while the percentage of women academics in STEM fields has risen dramatically over the past 20 years, as of 2013 women still made up less than 50 percent of academics at all levels, including non-tenure track instructors, and one-third or fewer of all tenure or tenure track faculty positions in core STEM fields.[25] Similarly, while the percentage of female full-time STEM professors more than doubled between 1993 and 2013, fewer than one-quarter of full professors in STEM fields at 4-year colleges and universities were women in 2013 (see fig. 6).[26]

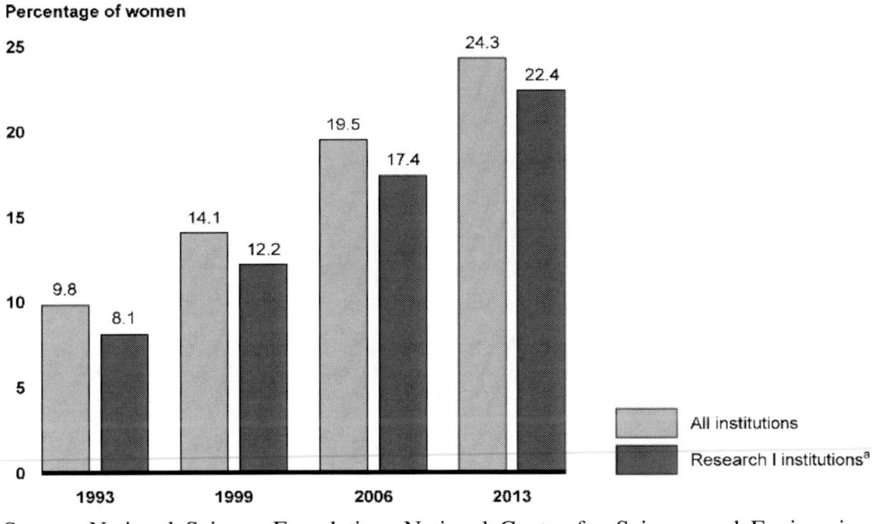

Source: National Science Foundation, National Center for Science and Engineering Statistics. | GAO-16-14.

Notes: Percentages in this figure include professors in health fields.

[a]According to the 1994 Carnegie classification, research I institutions are those that offer a full range of baccalaureate programs, are committed to graduate education through the doctorate level, award 50 or more doctoral degrees annually, and receive $40 million or more in federal research support annually.

Figure 6. Women as a Percentage of Full-time, Full Professors with STEM Doctorates, by Employing Institution, Fiscal Years 1993–2013.

The fact that women are underrepresented among doctorate holders in most STEM fields could affect the overall likelihood of women receiving federal funding. This is because the number of women choosing to study in STEM fields and pursue STEM careers—often referred to as the "pipeline"—greatly affects the number of female doctorates to whom agencies can potentially award research grants. There are numerous other factors that take place at the university level that may affect whether women choose to persist in STEM academic careers and eventually apply for federal research funding including:

- Lack of flexibility in academic career timelines: Studies show that having children is one of the primary reasons that some women trained in STEM disciplines choose not to remain in STEM research careers.[27] Many women in academia face the simultaneous demands of launching their research careers and beginning a family, and having children places pressure on women that men in STEM research do not experience.[28] Further, current academic promotion policies are based on rigid timelines and assume a total commitment—research has demonstrated that taking any time off to engage in caregiving responsibilities can be damaging to an individual's chances of receiving tenure.[29] Additional research has demonstrated that female post-doctoral researchers with no plans to have children go on to academic careers at the same rates as men, while female post-doctoral researchers who have children during their post-doctoral research appointment drop out of academia at twice the rate of their male counterparts.[30] Meanwhile, studies have found that the chances of obtaining tenure for men are not negatively affected by having children.[31]
- Lower propensity to apply for tenure and research grants: Research has demonstrated that women apply for tenure track faculty positions in disproportionately low numbers relative to the numbers of STEM doctorates they are granted, although conditional on applying, they are likely to succeed.[32] Research has also found that women apply for federal research grants at lower rates than their male counterparts. Factors such as family circumstances, lower self-confidence, and women leaving the research career path have been suggested as possible explanations for these lower application rates.[33]
- Unconscious biases: There is some evidence that both men and women have unconscious biases toward women in science.[34] In

> addition, characteristics such as "assertiveness" and "single-mindedness," traits research has shown to be typically associated with men and socially unacceptable for women, have been found to positively affect tenure decisions despite little evidence that they relate to scientific creativity.[35] Unconscious bias may contribute to an additional factor identified by researchers that may influence grant award success—the limited availability of mentoring for female STEM researchers and exclusion from informal academic networks.[36] Other studies have noted that subtle discrimination and departmental climate may be additional factors that discourage women from pursuing research careers.[37]

Previous research regarding differences between women and men in federal grant-making has been hampered by limited data. In 2005, RAND Corporation completed a study on behalf of NSF that examined federal grant funding at NIH, NSF, and USDA.[38] This report found that the probability of success after applying for federal grants was similar between female and male applicants across all three agencies. RAND Corporation was unable to conduct its analysis at DOD or DOE due to lack of data; it recommended these agencies improve their data collection. In our March 2015 report, we found that many of the same data limitations identified by RAND at some agencies persist 10 years later.[39]

Our Analysis Found No Disparities in Success Rates in STEM Research Grant Awards between Women and Men at Three Agencies, but Data Limitations Provide Limited Insights at the Other Three Agencies

Female and Male Grant Applicants Had Similar Success Rates at NIH, NSF, and USDA

At three of six agencies we reviewed—NIH, NSF, and USDA-NIFA—success rates for women and men were similar.[40] Based on 5 years of data from fiscal years 2009 to 2013, at NIH—the agency we reviewed with the greatest amount of federal research funding by far—the 5-year average success rates for women and men were the same. At NSF, the agency we reviewed with the second greatest amount of research funding, and USDA-NIFA, the 5-year average success rates for women and men were within 1 percentage point of each other. While average success rates vary slightly year by year at these three agencies, we found that the success rates for women and men were similar across all 5 years we analyzed and in some years, slightly higher for women. It is important to note that NIH, NSF, and USDA-NIFA receive fewer applications from women than from men and therefore make fewer grant awards to women. However, conditional on having submitted an application, women and men have similar chances of receiving funding from these agencies over the 5-year period we reviewed. Table 1 shows the 5-year average success rates for women and men at these agencies.

Table 1. Summary of Federal Grant Award Success Rates for Women and Men at Selected Agencies

Agency	Women[a] - Average Success Rate Fiscal Year 2009-2013	Men[a] - Average Success Rate Fiscal Year 2009-2013
National Institutes of Health	25%	25%
National Science Foundation	26%	25%
Department of Agriculture-National Institute of Food and Agriculture	22%	21%

Source: GAO analysis of agency data. | GAO-16-14

Notes: These percentages reflect the total population of proposals and awards for each agency and component and are not based on sample data. The figures presented are weighted averages based on pooled data across all fiscal years for which data were available (unless otherwise noted, we analyzed data from fiscal years 2009 through 2013). NIH, NSF, and USDA-NIFA received approximately 71,000, 46,000, and 2,000 applications per year, respectively, across all award categories over the five year period we reviewed.

[a]NIH, NSF, and USDA collect voluntary, self-reported data on the sex of applicants.

NIH, NSF, and USDA-NIFA all ask for and retain data on researcher qualifications and other individual characteristics that help these agencies describe the activities and analyze the performance of their grant-making programs.[41] As a result, these agencies are able to calculate detailed statistics on characteristics of their grantees, as well as detailed summary statistics on the number and type of proposals received, and amount of awards made.[42] In the NIH Office of Extramural Research's 2013 annual report, for example,

NIH explains that data are collected to provide NIH leadership with critical information for strategic decision making and to aid NIH's transparency and accountability to the public. For instance, in order to further the goal of sustaining a diverse biomedical workforce, the NIH Office of Extramural Research conducts extensive analysis of grant applicants, identifying risks and opportunities for improving participation of underrepresented groups.[43] NSF is required to report annually to the National Science Board on its merit review process, which includes publishing detailed statistics and analysis of their grant-making process.[44] In addition to reporting on success rates and information on grantee characteristics, NSF publishes statistics on the sex of reviewers, different types of peer review panels, and program officer characteristics. USDANIFA also collects and analyzes data and makes detailed information regarding its grant programs available on its website, citing transparency and internal evaluation as reasons for doing so.[45]

Data Limitations at DOD, DOE, and NASA Provide Limited Insights into Success Rates for Women and Men

For the components at DOD and DOE that had data we could analyze, differences in success rates between women and men varied.[46] DOD could only provide complete data for three of the eight components that we requested: the Army Research Office, the Defense Threat Reduction Agency, and the Office of Naval Research. These three components accounted for approximately half of DOD's STEM research awards to universities in FY 2014.[47] At the first two components we were able to analyze—the Army Research Office and the Defense Threat Reduction Agency—success rates between women and men were similar; that is, within 1 or 2 percentage points over the 5-year period we analyzed. At the third component—the Office of Naval Research—DOD's second largest research funding component—we found that men had a success rate that was 6 percentage points higher compared to women.

Table 2. Summary of Federal Grant Award Success Rates for Women and Men by Agency and Component at the Department of Defense (DOD), the Department of Energy (DOE), and NASA

Agency	Women-Average Success Rate Fiscal Year 2009-2013	Men-Average Success Rate Fiscal Year 2009-2013
Department of Defense		
Air Force Office of Scientific Research	Could not assess due to data limitations	
Air Force Research Labs[a]	Could not assess due to data limitations	
Army Research Office	48%	49%
Army Medical Command	Could not assess due to data limitations	
Basic Research Office	Could not assess due to data limitations	
Defense Advanced Research Projects Agency	Could not assess due to data limitations	
Defense Threat Reduction Agency	17%	17%
Office of Naval Research[b]	42%	48%
Department of Energy		
Advanced Research Projects Agency – Energy[c]	10%	7%
Energy Efficiency and Renewable Energy[d]	7%	5%
Nuclear Energy	Could not assess due to data limitations	
Office of Science	34%	41%
National Aeronautics and Space Administration	Could not assess due to data limitations	

Source: GAO analysis of agency administrative data. | GAO-16-14

Notes: The figures presented are weighted averages based on pooled data across all fiscal years for which data were available. (Unless otherwise noted, we analyzed data from 2009 through 2013, the most recent year of data available.) The approximate number of proposals received per year on average, are as follows: DOD-Army Research Office (900), DOD-Defense Threat Reduction Agency (300), DOD-Office of Naval Research (2,400), DOE-Advanced Research Projects Agency – Energy (500), DOE-Office of Energy Efficiency and Renewable Energy (400), DOE-Office of Science (2,700). These represent aggregate success rates for all award types that a given agency makes; for agencies that grant renewals, such as DOE's Office of Science, renewal success rates may be much higher than success rates for new awards, thus increasing the overall combined success rate. These percentages reflect the total population of proposals and awards for each agency and component and are not based on sample data. For this reason we do not present information about sampling error and any standard error information around our estimates. To test the accuracy, sensitivity, and robustness of this technique we tested our name-matching methodology against an agency dataset which included self-reported sex and also conducted additional robustness checks.

[a]The Air Force Office of Scientific Research is an office within the Air Force Research Labs, however it tracks information separately from Air Force Research Labs and therefore we analyzed it separately.

[b]We only analyzed 4 years of data (FY 2010 through 2013) from the Office of Naval Research due to agency-reported data reliability concerns with FY 2009 data.

[c]Only 4 years of data were available from the Advanced Research Projects Agency – Energy. In each year we analyzed, due to the small number of awards made (in FY 2010: 28; FY 2011: 30; FY 2012: 26; and in FY 2013: 22) these results are highly sensitive to the assumptions of our methodology.

[d]Only 3 years of data were available from the Office of Energy Efficiency and Renewable Energy. In each year we analyzed, due to the small number of awards made (in FY 2011: 27; in FY 2012: 16; and in FY 2013: 5) these results are highly sensitive to the assumptions of our methodology.

DOE could only provide complete data from three of four components that we requested: the Advanced Research Projects Agency-Energy, the Office of Energy Efficiency and Renewable Energy, and the Office of Science. These three components accounted for 85 percent of DOE's STEM research awards to universities in FY 2014. At two of DOE's smaller grant-making components— the Advanced Research Projects Agency-Energy and the Office of Energy Efficiency and Renewable Energy—we found success rates for women were higher than for men. These two components give out relatively few total awards and receive few applications from women each fiscal year, which means a difference of one or two awards in any given fiscal year can cause success rates to fluctuate substantially.[48] Meanwhile, data from DOE's largest grant-making component—the Office of Science—show that women had a 34 percent success rate, on average, over 5 years, while men had a 41 percent success rate over the same time period, a difference of 7 percentage points.

At NASA, we could not calculate success rates because we were unable to electronically link proposals to their corresponding awards in a systematic way or reliably identify which proposals came from researchers at educational institutions.[49] Our analysis of success rates at DOD, DOE, and NASA is summarized in Table 2.[50]

It is important to note that differences in success rates between women and men at DOD and DOE do not necessarily indicate discriminatory practices by federal funding agencies. Indeed, such differences could exist because women and men sometimes differ in qualifications—such as academic rank, discipline of doctorate, or years of experience—which can affect their chances of receiving a grant. We could not take these differences into account in our success rate analysis because most agencies did not collect data on researcher qualifications.[51] Instead, we used a separate dataset, NSF's 2013 Survey of Doctorate Recipients, to explore more generally whether differences in women and men's qualifications might explain some of the differences in the chances of receiving federal funding. Appendix II provides detailed information about our analysis of the Survey of Doctorate Recipients.

Current Data Collection Practices at NASA and Certain Components at DOD and DOE Hinder Oversight and Management of Federal Grant-making Programs

NASA, five of eight components we reviewed at DOD, and one component we reviewed at DOE do not collect and store data in such a way to enable them to systematically identify the characteristics of successful proposals and successful applicants, nor are they able to track statistics on funding success rates. As previously noted, we could not electronically match NASA's proposal data reliably to its award records in a systematic way, considering the volume of records, and therefore could not calculate a success rate for educational institutions. Several components at DOD and one component at DOE were unable to produce complete, accurate records of each grant's life cycle from proposal through award. For example, only three of eight components at DOD had usable historical data on both proposals and awards for the time period we reviewed.[52] One component at DOD was unable to provide us the name of the principal investigator associated with the grants they award, only tracking the name of the institution that employs the principal investigator. As we previously reported, three DOD components and one component at DOE do not retain proposal information at all.[53]

Additionally, at DOD and DOE, nearly all components told us that they require potential applicants to submit their research proposal ideas in writing, after which prospective applicants are either encouraged or discouraged from submitting a full proposal for formal consideration. However, some DOD and DOE components keep records of decisions regarding their pre-proposal process, while others do not, making it impossible for these components to determine whether this pre-proposal process has a differential effect on women versus men.[54]

In our previous report, we also noted that one reason complete life cycle grant data are difficult to obtain at some components is because fragmented data systems can make it difficult or even impossible to link records from different stages of the grant life cycle (e.g., linking a proposal record and its corresponding award record).[55] Additionally, some components at DOD and DOE reported that they do not currently collect certain demographic data containing personally identifiable information (PII) because they had no internal purpose for them or that they were not required by law to collect them, making it less likely that these agencies would maintain data systems that facilitate this type of analysis.[56]

According to officials at DOD, DOE, and NASA, they are not legally required to collect such data. However, because DOD, DOE, and NASA lack electronic records containing complete life cycle proposal and award data, they cannot systematically evaluate their programs' performance against their own stated program goals of funding the best science and the most qualified scientists, irrespective of gender. This lack of data hinders both internal management and external oversight. DOD, DOE, and NASA funding announcements all emphasize scientific merit and researcher qualifications in their stated criteria for research proposal review. Without collecting sufficient data on proposals and grants, however, these agencies cannot assess whether their award criteria are being applied consistently across their grant programs. Federal internal control standards also advise agencies to promptly record the entire process—or life cycle—of a transaction or event from initiation and authorization through its final classification in summary records. These standards also emphasize that internal controls should facilitate ongoing monitoring and evaluations, including agency self-assessments or external evaluations.[57]

Some agencies have indicated that they intend to take steps toward improving their data on proposals and awards. For example, in response to our March 2015 report, DOD's Office of the Under Secretary of Defense for Acquisition, Technology and Logistics (OUSD(AT&L)) reported on behalf of the Secretary of Defense that it would consult with other federal agencies that already collect this information to determine best practices before deciding on whether to expand its own data collection efforts. To date, however, OUSD(AT&L) has no documented plans or timeline in place for beginning this process. Also, we reported in March 2015 that NASA, through its Office of the Chief Scientist, was taking steps to pursue the feasibility of collecting additional demographic data (including age, race, and sex) about applicants to its competitive grant solicitations. As of the date of this report, NASA is currently in the process of obtaining approval from the Office of Management and Budget (OMB), in compliance with the Paperwork Reduction Act, to begin collecting these data. NASA has further stated that capturing and analyzing these data are critical steps it must take to advance its objectives, prohibit discrimination, and require equal opportunity in NASA-funded programs. At present, however, it is unclear if NASA plans to implement changes to its data systems that are needed to provide readily available, electronic, aggregate life cycle data on proposals and awards in a format appropriate for statistical analysis. Lastly, in 2015, one component at DOE—the Office of Science—began collecting demographic data on applicants on a voluntary basis, according to officials. However, the remaining three DOE components we reviewed do not collect this information and have no plans to do so at the present time.

Agencies Differ in How They Enforce Grantee Compliance with Title IX, and DOJ's Facilitation of Title IX Information Sharing Across STEM Agencies Is Limited

Only Four of Six Agencies Conduct Required Post-Award Compliance Reviews

Four of the six grant-making agencies we reviewed conduct periodic Title IX compliance reviews, as required by federal regulations.[58] Specifically, from fiscal year 2010 through 2014, DOE, NASA, NSF, and USDA-NIFA officials reported they found all grantees they reviewed to be in compliance with the basic requirements of Title IX.[59] Consistent with federal regulations, these four agencies each conduct periodic post-award compliance reviews of grantees to monitor Title IX compliance, recommend improvements to university policies and procedures, and identify promising practices to prevent sex discrimination on campus.[60] These grant-making agencies use criteria to select grantees for compliance reviews, including, but not limited to, the amount of funding the agency provides to the grantee, how recently a federal agency reviewed the grantee for Title IX compliance, and any suspected Title IX compliance issues at the grantee. Each of these four agencies primarily conducts compliance reviews onsite at the grantee's campus and uses several methods to gather information and determine compliance (see fig. 7).

In their written compliance reports, agencies may suggest a grantee take corrective action to improve existing efforts to prevent sex discrimination. For example, after one compliance review, DOE recommended that a grantee provide additional Title IX training to students, faculty, and staff, whereas after USDA-NIFA completed a compliance review, it recommended that a grantee review its hiring practices to ensure that there were no barriers to hiring women. NASA reported it has made recommendations for strengthening existing equal opportunity compliance efforts in every Title IX compliance report it has issued. These four agencies we reviewed typically allow grantees 1 to 6 months to implement any corrective actions they identify. During the compliance review process, these four agencies also identify promising practices to prohibit sex discrimination in university STEM departments.[61]

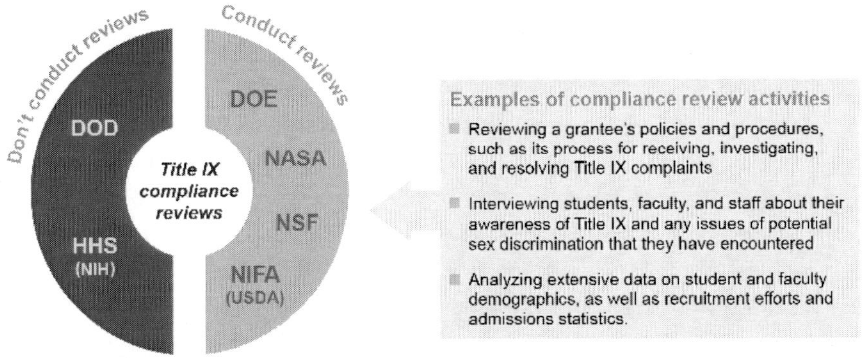

Source: GAO analysis of Title IX Compliance Review policies at Health and Human Services (HHS) National Institutes of Health (NIH); National Science Foundation (NSF); Department of Agriculture (USDA) National Institute of Food and Agriculture (NIFA); Department of Defense (DOD); Department of Energy (DOE); and NASA. | GAO-16-14.

Figure 7. Title IX Compliance Review Activities Taken by Six Selected STEM Grant-making Agencies.

We found that the remaining two agencies we reviewed, DOD and HHS—which is responsible for Title IX enforcement oversight of NIH's grantees—have not conducted required Title IX compliance reviews.[62] Officials at many of DOD's grant-making components told us they were unaware of the requirement to conduct compliance reviews.[63] Further, officials at DOD's grant-making components consistently mentioned a lack of departmental guidance on this responsibility and told us that they were not aware of any DOD-wide directive that instructs them to conduct periodic Title IX compliance reviews of their grantees. Officials at DOD's Office of Diversity Management and Equal Opportunity (ODMEO), the department-level office responsible for setting DOD-wide civil rights policy, told us that they are in the process of revising the Department's civil rights enforcement Directive pertinent to Title IX.[64] ODMEO officials said this revision will clarify the requirement and DOD components' responsibilities with respect to conducting periodic Title IX compliance reviews but were uncertain of whether this document would detail how the Department would ensure these reviews periodically take place. Officials further stated that this Directive is still being drafted and they did not know when it would be issued and disseminated to the Department's grant-making components. Meanwhile, all but one of the DOD grant-making components we reviewed currently take no action to enforce Title IX by conducting required periodic compliance reviews of their university grantees, thereby leaving a significant portion of federal STEM research funding to universities without potential Title IX enforcement oversight.[65]

With respect to NIH's grantees, HHS officials told us it has not conducted Title IX compliance reviews of NIH grantees in order to focus its civil rights enforcement efforts on a wider array of civil rights protections in health care settings, where it has primary or sole jurisdiction.[66] Officials also told us that one factor in the agency's decision not to focus on Title IX is that HHS has received only a few Title IX complaints against universities NIH funds for STEM research.[67] Furthermore, according to HHS officials, Education has the expertise to effectively enforce Title IX at universities, and HHS officials reported their agency has coordinated with Education on some Education-led compliance reviews that may involve NIH grantees. Education officials, however, told us they are unable to conduct compliance reviews on behalf of other federal agencies without a formal delegation agreement on file and said the agency does not currently have one in place with HHS. In addition, Education's current Title IX compliance activities are not focused on university STEM departments, according to Education officials. Although HHS does not conduct required Title IX compliance reviews of NIH grantees, HHS reported providing training to federal civil rights investigators across the government including information on Title IX enforcement responsibilities four times between 2012 and 2014.[68] While HHS's efforts to provide civil rights enforcement training to stakeholders may be helpful, it does not address the legal requirement to conduct compliance reviews. Given NIH's $15.5 billion in research funding provided to universities in fiscal year 2014[69]—more than 50 percent of all federal STEM research funding to universities in that year—the lack of a Title IX compliance review program at HHS leaves billions of dollars of NIH's STEM research funding to universities without potential federal Title IX enforcement oversight.

Existing DOJ Guidance Has Helped Several Agencies Establish their Title IX Compliance Review Programs, but Agencies Reported a Desire for More Information Sharing

DOJ has taken steps to coordinate Title IX enforcement across the federal government, including publishing two manuals—one that describes procedural considerations for conducting compliance reviews and another that outlines grant-making agencies' enforcement authority— but most agencies we reviewed reported a desire for DOJ to facilitate additional information sharing across agencies to continue to improve their Title IX compliance programs. In 1998, DOJ published an investigative procedures manual that describes specific steps and key considerations in the investigation and resolution of complaints filed against federal funding recipients under non-discrimination laws, including Title IX.[70] DOJ also published the Title IX common rule in 2000 and issued a Title IX legal manual in 2001, which, among other things, outlines agencies' enforcement authority.[71] Officials from several agencies we reviewed reported that their agency referenced these DOJ guidance documents and found them to be useful in developing their Title IX compliance review programs.

Despite the existing guidance provided by DOJ, officials from five of the six grant-making agencies we reviewed reported a desire for DOJ to facilitate greater information sharing. Specifically, several agency officials said they would benefit from sharing information with each other regarding best practices for compliance activities. Pursuant to Executive Order 12250, which was issued in 1980, DOJ is responsible for leading and coordinating efforts across federal agencies to enforce various federal non-discrimination laws, including Title IX. Specifically, the Executive Order states that DOJ shall "establish guidelines and standards for the...sharing and exchange by agencies of compliance records, findings, and supporting documentation."[72] In the absence of a DOJ-led process to share information on Title IX compliance efforts with STEM grant-making agencies, these agencies may not have effective opportunities to improve their compliance programs and coordinate with each other.[73] For example, officials from DOE, NASA, and NSF—three agencies with established compliance review programs—said that a DOJled information sharing effort would be helpful. For example, NSF officials said that it could be used for multiple agencies to come together to share Title IX compliance enforcement best practices, and NASA officials stated that it could be used to share more information on issues such as determining the scope of compliance reviews or strategies for federal agencies to enforce compliance. In addition, such information sharing could assist grant-making agencies with challenges they face. For example, NASA officials said they were in need of additional information sharing to discuss specific elements of compliance reviews, including certain subject areas covered by Title IX such as recruitment and counseling. Further, without additional information sharing, agencies that currently do not conduct compliance reviews may not be aware of resources or other agency expertise available that could help them establish compliance review programs. For example, a civil rights official from DOD said that civil rights officials across the Department generally lacked awareness of their Title IX compliance responsibilities. She said that DOJ could help improve awareness at DOD by convening periodic meetings for federal funding agencies to learn about their Title IX compliance responsibilities and share information with each other.

At present, DOJ has no formal process for facilitating information sharing about Title IX compliance activities with regard to STEM grantees across these six STEM grant-making agencies. From 2005 to 2010, DOJ led a Title IX STEM initiative, including facilitating a Title IX interagency working group comprised of DOE, NASA, and NSF. Officials from these agencies reported the working group was extremely helpful to their compliance review efforts. However, this group no longer exists and DOJ instead provides STEM-related Title IX technical assistance to agencies that request it.[74] DOJ cited resource constraints as one impediment to undertaking additional activities related to Title IX.[75] While responding to requests for information from individual agencies can be helpful, coordinating information sharing across agencies could also be an efficient and cost-effective approach for DOJ to meet the information needs of several agencies at once while reducing the resources DOJ must expend to respond to individual agency inquiries. Moreover, a more unified and inclusive approach to information sharing by DOJ could also raise awareness among a broader range of agencies about their Title IX compliance responsibilities by bringing together agencies with a range of Title IX compliance experience to learn from each other. DOJ officials said they recognize the need to develop more efficient, cost-effective ways to coordinate with federal agencies and reported that they are in the process of assessing their activities for Title IX.

Through a Literature Review and Expert Consultation, We Identified 13 Actions That Federal Agencies May Choose to Take to Help Address Women's Representation in Federal STEM Research

Through our literature review and consultation with experts, we identified 13 possible actions for grant-making agencies to consider that could help address women's representation in STEM research.[76] The actions we identified were focused both internally within the agency and externally, with respect to an agency's grantees. We grouped these actions into four separate categories: (1) enhancing agency leadership and collaboration, (2) establishing family-friendly policies for grantees, (3) overseeing the research proposal review process, and (4) funding and assisting academic institutions. While some of these actions do not specifically target women, they could help address key challenges researchers have found that women face in STEM research careers, such as balancing career and family responsibilities, working in unsupportive environments, and having limited access to career development resources.[77] Figure 8 presents the four categories of possible actions we identified, and the following sections describe each category in more detail.[78]

Source: GAO analysis of literature, expert input and federal agency actions. | GAO-16-14.

Notes: The actions listed under each category above are abridged versions of the actions we sent to the experts and agencies. Please see appendix III for a full description of each action.

Figure 8. Categories of Actions GAO Identified through a Literature Review and Expert Consultation That Federal Agencies Could Consider to Help Address Women's Representation in Federal STEM Research.

All six agencies we reviewed are taking actions in some of these areas, though at the same time, most agencies also reported potential challenges in implementing some of these actions. Specifically, certain agencies' missions may limit the relevance or feasibility of some of the actions. For example, DOD reported that its mission limits its authority to fund research beyond national security related projects, making it infeasible for it to fund research on cultural and structural barriers to women's success in STEM fields. Nonetheless, officials at most agencies we reviewed reported a willingness to explore the feasibility of implementing several of these actions.

Enhancing Agency Leadership and Collaboration

Experts we consulted agreed a number of actions related to enhancing agency leadership and agency-wide collaboration could help aid in the recruitment, retention, and advancement of women in STEM research. Studies have shown that women leave STEM fields at a higher rate than their male peers. For example, one study found that women leave STEM academic positions at higher rates than men, in part due to dissatisfaction with departmental culture, faculty leadership, and research support.[79] In addition, our prior work has found that a commitment from top leadership is an important component of cultivating a supportive work environment.[80] To address these needs, 14 of 19 experts we consulted agreed that agencies could establish a working group that reports directly to the agency head and has agency-wide scope to not only assess successful strategies, but also develop programs and policies that aid in the recruitment, retention, and advancement of women in STEM research. For example, NIH established an agency-wide working group focused on women in biomedical careers to develop institutional programs and aid the recruitment, retention, and advancement of women in biomedical faculty and leadership positions. According to one expert we consulted, such groups provide a forum for agency leaders to gather data on their agency's performance and develop strategies to improve their agency's performance. Other experts noted that conditions vary among different STEM disciplines, and one suggested that disciplinary experts could bring knowledge of the nuances of each field to the agency-level working groups.

Research has also shown that women scientists lack guidance on how to advance in their academic careers while also balancing work and family.[81] Of the experts we consulted, 13 of 19 agreed that agencies could maintain publicly-available websites that provide guidance for women in STEM research careers, including information on mentoring and career development opportunities. For example, NSF launched its Career Life Balance Initiative website to help researchers navigate STEM career-life pathways from graduate education to full professorship. In addition, NASA and DOE maintain websites with biographies on women in science, and NIH supports a social media website for women of color in biomedical careers and for other supporters of diversity in the scientific workforce. This website provides information on mentoring and career development opportunities.

Establishing Family-Friendly Policies for Grantees

Experts we consulted agreed that a number of targeted family-friendly supports could help researchers balance their work and family responsibilities, a challenge which studies show disproportionately affects women.[82] For example, one study found that women in the sciences who were married with children were 35 percent less likely to enter a tenure track position after receiving their PhD than married men with children.[83] Of the experts we consulted, 14 of 19 agreed that providing support for researchers to transition back to their careers after taking family leave could help improve the representation of women in STEM research. For example, the NIH Reentry into Biomedical Research Careers program supplements existing NIH research grants to support full- or part-time research by women or men returning to the scientific workforce. The program is designed to bring a scientist's existing research skills and knowledge up-to-date so that by end of the supplement period, the scientist will be prepared to apply for an NIH career development award, an NIH research award, or another form of independent research support.

Research suggests that childcare support can also help address challenges related to balancing work and family that disproportionately affect women.[84] Of the experts we consulted, 15 of 19 agreed that covering certain childcare costs could help support researchers with children. For example, NIH officials reported that most NIH grant awards allow for reimbursement of certain childcare costs.[85] In addition, in 2010, NIH implemented a policy that requires applications for conference support to include descriptions of child and family care offerings at the conference site.

Overseeing the Research Proposal Review Process

Experts we consulted also supported actions related to mitigating the potential for bias in the research proposal review process. Research on job selection processes has documented cases of gender bias, including in STEM fields. One study found that science faculty demonstrated subtle gender bias in favor of male applicants when evaluating fictitious applications for research positions in laboratories.[86] Other research has shown that temporarily hiding the identities of job applicants during early stages of the application review process—a partially blind review—helped women and minorities make it to the interview stage, and others suggest it could also help women's odds of acceptance for scientific papers and grant proposals.[87] Although most of the federal funding agencies we reviewed do not conduct blind reviews of research proposals, DOE's Office of Nuclear Energy has been conducting partially blind reviews since 2010. In their review, grant applicants' identities are not revealed until after the panel has assessed and scored the applicants' initial technical proposals. Nuclear Energy officials reported this process has been successful in diversifying its pool of grantees to include younger researchers, female researchers, and researchers new to the Office of Nuclear Energy.

Research has found that diversity brings different information, opinions, and perspectives to a group, leading to better decision-making than non-diverse groups.[88] Of the experts we consulted, 14 of 19 supported efforts to recruit a diverse set of proposal reviewers to broaden the participation of women in STEM research. For example, NSF reported that it has initiated a 3-year pilot effort – a virtual panelist project – to enable proposal reviewers to participate remotely in review panels in order to reduce barriers to participation and plans to review the impact of the use of virtual review panels.

Funding and Assisting Academic Institutions

Experts agreed that several actions related to funding additional academic research and providing technical assistance to universities could further knowledge about the obstacles women face in STEM research fields and determine possible interventions. Studies we reviewed that examine possible obstacles women face in STEM fields have primarily focused on biomedical fields, and only limited research has been conducted to test and evaluate possible interventions.[89] Of the experts we consulted, 16 of 19 supported conducting additional research on the obstacles to diversity in STEM research fields, including exploring the differential causes of women's underrepresentation in academia by STEM discipline and evaluating university-level interventions that have been proposed to increase women's participation in STEM research. For example, NSF provides research funding to universities through its ADVANCE program to address various aspects of STEM academic culture and institutional structure that may differently affect women faculty and academic administrators. Additionally, NIH awarded 14 grants in 2009—totaling $16.8 million over 4 years—to further research on the causal factors and interventions that promote and support the careers of women in biomedical and behavioral science and engineering. It also hosted workshops to present research findings and discuss possible interventions.

Of the experts we consulted, 11 of 19 agreed that encouraging grant-recipient institutions to conduct unconscious bias workshops could also improve women's representation in STEM. For example, NASA developed an online unconscious bias training tool for universities to use. Though experts we consulted generally agreed that unconscious bias is a problem, several cautioned that any unconscious bias training should be carefully designed, implemented, and evaluated to ensure its effectiveness.

CONCLUSION

Each year, multiple federal agencies spend billions of dollars funding STEM research—an area in which research shows women continue to be underrepresented. Several federal grant-making agencies have elected to take steps to help women access and remain in STEM research. However, as long as NASA and some components at DOD and DOE elect not to retain complete data on their grant-making process in an analyzable, electronic format, they will not fully understand how well they are achieving their goals of funding the best science and most qualified scientists, irrespective of gender. Without collecting demographic data on applicants, they will also be unable to analyze how women in particular fare in their grant programs. Lack of these data will continue to hinder their ability to make informed decisions about how to improve their grant-making programs. Moreover, if complete data regarding the federal grant decision-making process continues to be unavailable for both internal evaluations and external oversight, it will be difficult to hold these agencies accountable for their decisions regarding the allocation of billions of research dollars.

DOJ has developed some guidance to help agencies conduct Title IX compliance reviews, but because there is not a process to facilitate information sharing about such reviews of STEM grant recipients across relevant agencies, it misses a valuable opportunity to help STEM grant-making agencies enhance their compliance review programs. All federal agencies that fund STEM research at universities are responsible for enforcing Title IX requirements at their grantee institutions. In failing to conduct required post-award compliance reviews of DOD and NIH grantees, however, DOD and HHS are missing opportunities to identify potential weaknesses, improve procedures, and build institutional capacity so their university grantees may enforce and uphold a nondiscriminatory environment. Consequently, billions of dollars of federal research funding to universities could be left more vulnerable to discriminatory practices.

RECOMMENDATIONS FOR EXECUTIVE ACTION

In order to ensure complete, analyzable records regarding research grant award decisions are available for management and analysis, we recommend that the Secretary of Defense direct the Under Secretary of Defense for Acquisition, Technology and Logistics to lead the implementation of additional data collection efforts in coordination with DOD's grant-making components. These should include:

- Retaining complete records of pre-proposal, proposal, and award data, including a record of proposal disposition, in linked electronic files to facilitate aggregate, statistical analysis of the grant-making process, including the calculation of success rates.
- Collecting demographic, education, and career information from applicants, on a voluntary basis, that is not available to proposal reviewers but is used for analysis of success rates.

In order to ensure complete, analyzable records regarding research grant award decisions are available for management and analysis, we recommend that the Secretary of Energy direct DOE's grant-making agencies to implement additional data collection efforts, which should include:

- Retaining complete records of pre-proposal, proposal, and award data, including a record of proposal disposition, in linked electronic files to facilitate aggregate, statistical analysis of the grant-making process, including the calculation of success rates.
- Collecting demographic, education, and career information from applicants, on a voluntary basis, that is not available to proposal reviewers but is used for analysis of success rates.

As NASA begins to collect demographic data on its grant proposals and awards, we recommend the NASA Administrator include the following key components:

- Retain complete records of pre-proposal, proposal, and award data, including a record of proposal disposition, in linked electronic files to facilitate aggregate, statistical analysis of the grant-making process, including the calculation of success rates.
- Collect demographic, education, and career information from applicants, on a voluntary basis, that is not available to proposal reviewers but is used for analysis of success rates.

To improve Title IX enforcement by federal STEM grant-making agencies, we recommend the Principal Deputy Assistant Attorney General for the DOJ Civil Rights Division establish a process to facilitate information sharing across federal STEM grant-making agencies regarding current Title IX compliance efforts to promote equitable access to STEM research funds.

To comply with Title IX enforcement requirements, we recommend the Secretary of the Department of Defense, which funds STEM research at universities, direct the Director of the Office of Diversity Management and Equal Opportunity to ensure that Title IX compliance reviews of DOD's grantees are periodically conducted.

To comply with Title IX enforcement requirements, we recommend the Secretary of the Department of Health and Human Services, which funds STEM research at universities, ensure that Title IX compliance reviews of NIH's grantees are periodically conducted.

AGENCY COMMENTS AND OUR EVALUATION

We provided a draft of this report to the departments of Agriculture, Defense, Education, Energy, Health and Human Services, Justice, the National Aeronautics and Space Administration, the National Science Foundation, and the Office of Science and Technology Policy for review and comment. We received written comments on a draft of this report from the Departments of Defense, Energy, Health and Human Services, Justice, and the National Aeronautics and Space Administration. We also received technical comments from the Departments of Energy, Health and Human Services, Justice, and the

National Aeronautics and Space Administration, which we incorporated where appropriate.

DOD concurred with our recommendations that it implement additional data collection efforts and ensure that periodic Title IX compliance reviews of DOD grantees are conducted. DOD noted that it has updated its instructions on compliance with Title IX and other civil rights compliance and that these are in the process of review and approval for publication in the Federal Register.

DOE generally agreed with our recommendation that it implement additional data collection efforts but commented on three aspects of our review. First, DOE commented on the limitations inherent in our use of the Social Security Administration's baby names database in calculating DOE's rate of awarding STEM research grants to women. We would like to reiterate, however, that this technique has not only been used in previous peer-reviewed studies, but we also conducted sensitivity analyses that provided us with reasonable assurances of the validity of this technique. After careful consideration, we decided this was the best systematic, data-driven technique for calculating success rates at agencies—including DOE—that do not collect self-reported data on applicants' sex. As DOE points out, the limitations of this technique— which we acknowledge in detail in our report—highlight the need for additional data collection. If DOE were to collect additional data, it would be able to carry out improved empirical analyses in the future. Second, DOE cited concerns about our analysis of data from NSF's Survey of Doctorate Recipients (SDR), which we used in a separate analysis in our report. It is important to note that we did not use the SDR analysis to draw conclusions about DOE's rate of awarding STEM research grants to women. Instead, we used the SDR analysis as a supplement to our success rate calculation to analyze the relationship between overall researcher demographic, educational, and career characteristics and the chances a researcher receives funding from DOE. If DOE begins to collect more complete administrative data on its researchers' demographic, educational, and career characteristics, the use of external data sources like the SDR will not be necessary for this type of analysis. Lastly, DOE clarified that its legal concerns regarding the collection of data pertained only to personally identifiable information. We modified the report accordingly.

HHS agreed with our recommendation to initiate periodic Title IX compliance reviews. In addition, HHS detailed its broad approach to its Title IX compliance activities. We note these activities in our report and acknowledge they may serve a useful purpose for HHS-funded universities and other partners. As HHS implements our recommendation, it will be important for the agency to take additional measures and ensure that universities that receive STEM research grants from NIH are subject to periodic Title IX compliance reviews.

DOJ generally agreed with our recommendation that it facilitate additional information sharing among STEM grant-making agencies regarding Title IX compliance activities and provided additional information on its current efforts and role in information sharing. First, DOJ noted that it already engages in several coordination activities relevant to Title IX generally, as we documented in our report. Second, DOJ expressed concern that the draft report could give readers the impression that federal agencies would not be able to meet their duty to conduct periodic Title IX compliance reviews without additional information sharing by DOJ. We agree that federal agencies have a legal obligation to conduct periodic Title IX compliance reviews and believe that the report clearly describes these obligations and that our recommendations reemphasize this responsibility at the agency level. Third, DOJ agreed to take a leadership role in enhancing agencies' opportunities to share Title IX compliance information, but stated that Executive Order 12250 does not require it to do so. However, Executive Order 12250, entitled Leadership and Coordination of Nondiscrimination Laws, states that DOJ "shall coordinate the implementation and enforcement by Executive agencies of various nondiscrimination provisions" including Title IX, and "shall establish guidelines and standards for...the sharing and exchange by agencies of compliance records, findings, and supporting documentation..." As such, we believe that DOJ's leadership and coordination role with regard to Title IX under Executive Order 12250 puts it in a unique position to do more to engage in information sharing with STEM grant-making agencies to identify potential weaknesses, improve procedures, and build institutional capacity to strengthen their compliance review efforts. In its response, DOJ also described some of the ways it is considering responding to our recommendation, including holding quarterly teleconferences among the six agencies included in our review and hosting a meeting of the former Title IX STEM Interagency Working Group to discuss Title IX compliance issues. We believe either of these efforts would be an appropriate way for DOJ to formalize its information sharing among these six STEM grant-making agencies.

NASA partially concurred with our recommendation and noted several actions the agency is either considering or in the process of implementing as it prepares to collect basic demographic data on its grant applicants beginning in

2016. However, NASA commented on several challenges it faces in implementing aspects of our recommendation. First, with regard to retaining information received from applicants prior to submitting a full proposal—which we term the "pre-proposal" stage—NASA stated that pre-proposal data are the intellectual property of proposing institutions and that no data are received prior to the submission of a proposal. By "pre-proposal" we are not referring to the intellectual property of individual researchers and institutions, but rather to any information that is voluntarily submitted to NASA by the applicant in the form of white papers, abstracts, letters of intent, or other preliminary materials NASA may use to either informally or formally screen applicants prior to the next stage in the review process. It is our understanding that some of NASA's funding opportunities—similar to other agencies we reviewed—may ask for this preliminary information to provide potential applicants with feedback on their research ideas and in some cases to either encourage or discourage them from submitting a full proposal for funding. We maintain, therefore, that NASA should collect data on all phases of the grant review process, including the "pre-proposal" stage, where applicable, to support the aggregate statistical analysis of its grant-making process and calculation of success rates. Second, NASA stated that OMB is currently reviewing its request for authorization to collect basic demographic data and that modifying this request to include education and career information would delay the agency's efforts. NASA said it would prefer to continue with its current effort to collect data and then reassess the results in this regard. NASA said it could do a gap analysis to capture additional education and career information. We support such a gap analysis and believe these additional data could be collected separately from the request undergoing OMB review. As NASA noted, it already collects a great deal of educational and career information from applicants through resumes or CVs submitted during its application process. We encourage NASA to find a way to leverage this additional information to facilitate aggregate statistical analysis of its grant-making process and calculation of success rates.

Lastly, NASA expressed concern for applicants' loss of anonymity should demographic, education, and career information be collected. We agree that any personally identifiable information provided by applicants is confidential and should be appropriately safeguarded. NSF, along with other agencies we reviewed, already collect voluntary demographic, education, and career data from its grant applicants. These agencies have established processes for keeping this information both confidential and, in the case of demographic data, separate from the proposal, which helps to eliminate any potential for this data to inadvertently affect the award selection process. We believe NASA could establish similarly secure processes to protect this information.

Melissa Emrey-Arras
Director, Education, Workforce,
and Income Security

APPENDIX I: OBJECTIVE, SCOPE, AND METHODOLOGY

We addressed three objectives in this report: (1) the extent to which differences exist in federal research grant awards between women and men in STEM fields and what factors might explain any differences, (2) the extent to which federal agencies enforce Title IX at universities they fund for STEM research, and (3) possible actions federal agencies could take to address the representation of women in STEM research. This appendix provides details of the data sources used to answer these questions, the analyses we conducted, and any limitations we encountered.

We included in our review the following six federal STEM research grant-making agencies:[1]

- Department of Agriculture-National Institute of Food and Agriculture[2] (USDA-NIFA),
- Department of Defense (DOD),
- Department of Energy (DOE),
- National Aeronautics and Space Administration (NASA),
- National Institutes of Health[3] (NIH), and
- National Science Foundation (NSF).

We met with officials at each agency and asked them to identify the programs and components within their agency that fund grants to and cooperative agreements with researchers in STEM fields at educational institutions.[4] We excluded contracts and awards to non-educational institutions from this review. Table 3, below, shows the complete list of agencies and components included in our review.

Table 3. Agencies and Components in Our Review

Agency	Agency Component
Department of Agriculture (USDA)	National Institute of Food and Agriculture (NIFA)
Department of Defense (DOD)	Air Force Office of Scientific Research
	Air Force Research Laboratory
	Army Medical Command
	Army Research Office
	Basic Research –Office of Assistant Secretary of Defense for Research and Engineering
	Defense Advanced Research Projects Agency
	Defense Threat Reduction Agency
	Office of Naval Research
Department of Energy (DOE)	Advanced Research Projects Agency-Energy
	Office of Energy Efficiency and Renewable Energy
	Nuclear Energy
	Office of Science
National Aeronautics and Space Administration (NASA)	
National Institutes of Health (NIH)	
National Science Foundation (NSF)	

Source: GAO. | GAO-16-14

OBJECTIVE 1 METHODS

To determine if differences existed in federal grant-making to women and men in STEM fields, we obtained available administrative data from each agency included in our review for all research grant proposals received and awards made in fiscal years 2009 through 2013, the most recent complete year of agency data available when we began our review. We evaluated each agency's data collection practices by assessing the extent to which the data collected would allow agencies to evaluate their program's performance against their own stated program goals. We also compared their data collection practices against federal internal control standards, which outline a set of key elements of record keeping which agencies should have in place in order to ensure effective program management and also to allow for internal and external evaluation and oversight.[5] To complement our analysis of the agency administrative data and identify factors that may contribute to individual's success obtaining federal funding, we analyzed data from the National Science Foundation's Survey of Doctorate Recipients (SDR) from 2013.[6] We analyzed data from these two sources because neither data source alone could fully answer our research question; agency administrative data provided insight on the extent to which differences exist in federal STEM research awards and SDR data provided insight on what factors may explain any differences. First, we used the agency administrative data to analyze proposals and awards and to develop a set of descriptive statistics regarding federal grant funding to women and men. However, due to data limitations, we could not use agency administrative data to control for factors that might explain those numbers or otherwise explain how female and male awardees are similar or different because not all of the agencies and components collect the data necessary to examine and control for researcher characteristics that might be correlated with grant success.[7]

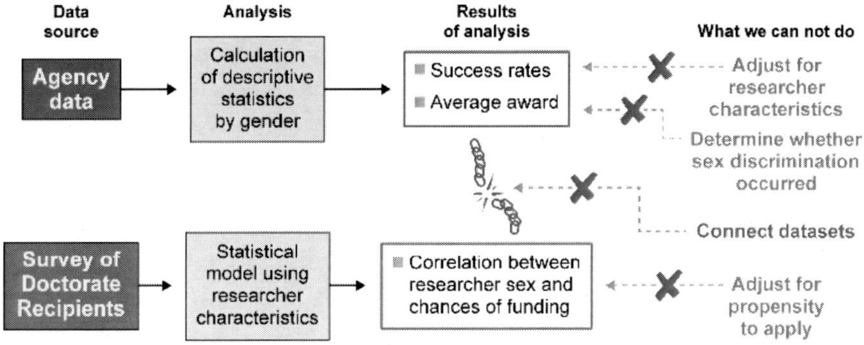

Source: GAO. | GAO-16-14.

Figure 9. How Data Sources Are Used In This Report.

Because of these limitations to analyzing the agency data alone, we also analyzed the SDR data to compare the odds of being funded for female and male STEM doctorates with similar characteristics at four grant-making agencies (NIH, NSF, DOE, and DOD) about which the SDR collects funding information.[8] Using the SDR, we examined the relationship between researcher characteristics, in particular the researcher's sex, and the chances that they are funded by the federal government or by one of these four agencies.

It is important to note that these analyses were not designed to and cannot be used to determine whether sex discrimination exists at the agencies we reviewed. Specifically, we could not link the administrative agency data with records from the SDR data to test whether success rates differed between similarly qualified women and men. That said, where the analysis points to unequal funding outcomes for similarly qualified men and women, more data would be needed to explain the source of these differences and it is possible that these differences would persist even with better data. Figure 9 illustrates how we used each of these data sources in our analysis.

Data Reliability

Through consultations with agency officials responsible for the data systems and by reviewing agency datasets, we assessed the reliability of all the data we present in this report and included in our analysis only those data that we found to be sufficiently reliable for the purposes of our report. We conducted data reliability testing on the agency datasets in a number of ways including reviewing agency documentation, interviewing agency officials responsible for data collection and storage, sending questionnaires to agency data analysts and database administrators, as well as electronic testing and inspection of the data. Through this process, we found that while some agencies did collect proposal and award data, it was not sufficiently complete or reliable for us to use in this analysis. For example, one agency we reviewed maintains proposals and awards in separate databases. We realized upon inspecting the data and attempting to match proposals to their corresponding awards that these data elements could not be electronically matched at the aggregate level—so success rates for a given fiscal year could not be calculated. Furthermore, this same agency did not consistently track institution type (electronic testing revealed that the field was not populated for the majority of records) so we were unable to reliably identify grants to only educational institutions. In the case of another agency, date information and proposal amount were only retained for successful proposals, while electronic testing and visual inspection of the data revealed these fields were not populated for unsuccessful proposals. In such cases, we excluded these datasets from our success rate analysis. Data we assessed from some components at DOD and DOE (specifically Advanced Research Projects Agency-Energy, Office of Energy Efficiency and Renewable Energy, and Office of Science at DOE, and Army Research Office, Defense Threat Reduction Agency, and Office of Naval Research at DOD) were determined to be sufficiently reliable for our purposes in this report.

For summary statistics provided by NIH and NSF, we reviewed documentation regarding the sources of the data, sent questionnaires to agency officials responsible for generating the information, and, in some cases, reviewed the code used to generate the summary statistics. We determined that the summary data provided by NIH and NSF were sufficiently reliable for our purposes. We also reviewed documentation for both the Survey of Doctorate Recipients data and the Social Security Administration name database and determined them to be sufficiently reliable for our purposes in this report.

Table 4. Researcher Information Collected by Selected Agency Administrative Data Systems

Variables collected	U.S. Dept. of Agriculture[a]	Dept. of Defense[b]	Dept. of Energy[c]	NASA	National Institutes of Health	National Science Foundation
Sex	X				X	X
If tracked, approx. % of records with sex recorded[d]	86%				90-95%	85-90%
Race	X				X	X
Age/year of degree	X	1 of 8 components			X	X
Institution	X	X	X	X	X	X
Highest degree	X	2 of 8 components			X	X
Discipline (or subject area of grant)	X	4 of 8 components	3 of 4 components	X	X	
Co-Investigator information[e]	X	2 of 8 components		X	X	X

Source: GAO analysis of agency documents. | GAO-16-14

Notes: This table has been updated since published in our prior report (GAO-15-291R). Specifically, we have included a checkmark in the co-investigator row for NSF.

[a]The only component included in our review at USDA was the National Institute of Food and Agriculture (NIFA).

[b]Three of eight components included in this review at DOD do not retain proposal data for grant applicants once awards are made.

[c]One of four components included in this review at DOE does not retain proposal data for grant applicants once awards are made.

[d]Demographic information such as sex and race are collected on a voluntary basis only; applicants are not required to provide this information to the agency. The percentages in this table were reported to us by the agencies. NIH and NSF provided us a range based on the past several fiscal years and USDA-NIFA provided us the percentage from their most recent full year of data, FY 2013.

[e]In this row, agencies received a checkmark if they collected any information regarding co-investigators, such as name or institution.

Calculating Agency-Level Success Rates for Grant Awards

To obtain the agency administrative data, we first asked each agency and component we reviewed to provide information regarding the data they collect on research grant proposals and awards and the data systems they use to store and manage these data. In March 2015, we reported on the data elements federal STEM grant-making agencies collect, and the data systems at these agencies.[9] Table 4 is an updated summary of the data elements each agency collects that would allow one to analyze success rates and researcher characteristics that could factor into any differences in success rates.

Using this information, we asked each agency or component to provide us with data on all proposals and awards they made in STEM fields to researchers at universities from 2009 through 2013, the most recent data available at the time we began our review.[10] We requested data from each agency or component that would allow us to calculate success rates, i.e. the ratio of awards made to proposals reviewed (specifically, those proposals with a final disposition), for women and men. We also collected additional information to assess the feasibility of analyzing the ratio of amount awarded to amount requested for women and men.[11] Specifically, for each proposal and award record, the key data elements we requested were:

- Amount awarded
- Amount requested
- Award status (approved, declined)
- Award type
- Award unique ID, link to unique proposal ID
- Start date of award
- End date of award
- Fiscal Year of award
- Institution
- Institution type (college vs. research university)
- Proposal unique ID
- Principal investigator first name
- Sex (where available)
- Type of proposal (reapply, modification, initial)
- Amount awarded for each corresponding award, sex
- Amount requested in each proposal, sex
- Award type, unique award ID

We analyzed data from each agency and component separately since data collection practices varied widely, even within agencies. To analyze agency success rates, we carried out three basic steps:

1. Data cleaning and filtering to isolate proposals and awards related to extramural grants and cooperative agreements to principal investigators working in STEM research at educational institutions only.

2. Where data on researcher sex were not available, we matched names to a given gender based on whether a name was given to a male or female at least 95 percent of the time in the Social Security Administration (SSA) baby name database between 1933 and 1993.

3. We calculated summary statistics overall and by the categories male, female, and unassigned. These summary statistics included:

a. Number of proposals

b. Number of awards

c. Success rate—the number of awards divided by the number of proposals.[12]

Each agency may have different overall success rates depending on their budgets, the number and sizes of awards they wish to make, their use of pre-proposals, and the number and types of applications received. The relevant comparison for the purposes of this report is not the difference in success rates between agencies, but whether or not there are differences in success rates between women and men within an agency. This statistic controls for the fact that fewer women than men apply for awards from federal agencies and instead focuses on chances of success conditional on having already applied. For example, if the success rate is 25 percent, that means that for every 100 grant applications received, 25 were ultimately awarded a grant. If success rates at an agency are equal for women and men, it implies that once women and men apply for a grant from that agency, their chances of being funded are the same on average.

Gender Name Match Methodology

Only three agencies in our review—NIH, NSF, and USDA—collect the self-reported sex of researchers that enabled us to easily calculate success rates. For the remaining agencies and components in our review, we requested the first name of the principal investigator applying for the grant.[13] We then assigned this first name to either 'female' or 'male' using a gender name match algorithm and the Social Security Administration's database of baby names (which provides data on all names given to infants applying for Social Security numbers in a given year and the sex of those infants).[14] We tested various thresholds for assigning a name to a given gender. Ultimately, we used a 95 percent threshold for assigning a name as either 'male' or 'female' based on the number of occurrences in the Social Security Administration's database. That is, we considered a grant applicant 'female' if that name was given to a girl baby 95 percent or more of the time, and 'male' if the name was assigned to a boy baby 95 percent or more of the time, according to the SSA.[15] While on average this should result in a relatively low error rate, one or two incorrectly assigned names at an agency that only gives out 5 awards in a year could materially affect the success rate calculation for women and men in the way that it would not at an agency that gives out hundreds or thousands of awards each year. For this reason, we make note of instances where success rates are calculated on a small number of records.

Table 5. Gender Name Match Percentages for Grant Proposal and Award Data at Department of Energy (DOE) and Department of Defense (DOD) Components

	Total Number of Records	Name assigned female	Name assigned male	No assignment
DOD-Army Research Office	**4567**	10%	62%	29%
DOD-Defense Threat Reduction Agency	**1560**	11%	62%	26%
DOD-Office of Naval Research	**9413**	12%	64%	24%
DOE-Advanced Research Projects Agency-Energy	**2078**	7%	61%	32%
DOE-Energy Efficiency and Renewable Energy	**1213**	10%	63%	27%
DOE-Office of Science	**13651**	13%	63%	25%

Source: GAO. | GAO-16-14

Notes: These numbers represent the percentage of names matched in each category using a 95% sex threshold; i.e. if a name was given to a male 95% of the time in the Social Security Administration (SSA) name database, the name was classified as male; if a name was given to a female 95% of the time in the SSA database, the name was classified as female. Numbers in a given row may not add up to 100% due to rounding.

Using this approach, we were able to match between 68 and 76 percent of the names in the data from DOD and DOE components to a given gender.[16] Table 5 shows the percentage of names we were able to match at specific DOD and DOE components for which we had data.

Sensitivity Tests of Gender Name Match Methodology

We conducted two sets of sensitivity tests and concluded the gender name match methodology was sufficiently reliable for the purposes of our analysis. First, we conducted a robustness check of our match by comparing the results of our match to the self-reported sex of applicants in USDA-NIFA's data. For approximately 20 percent of the names, our algorithm was unable to assign a gender. In approximately 1 percent of cases our algorithm assigned a gender to a name that directly conflicted with a person's self-reported sex. Table 6 shows the comparison of our gender name match to the self-reported data in the USDA-NIFA database.

Table 6. Match Performance versus Self-reported Data in Department of Agriculture-National Institute of Food and Agriculture (USDA-NIFA) Proposal and Award Database

	Match rates	Total number of records
GAO gender name match exactly the same as self-reported male/female sex	68%	**6709**
GAO gender name match assigned gender to a name where self-reported gender was missing or not provided	11%	**1131**
GAO gender match directly conflicted with self-reported male/female sex	1%	**101**
GAO could not assign a match to a name with missing or self-reported male/female sex	20%	**2000**

Source: GAO. | GAO-16-14

Second, we conducted a robustness check of our results that tested the limitations of our gender name match by assuming that all unassigned records were either 100 percent male or 100 percent female, respectively (see table 7). For example, at DOD-Office of Naval Research and DOE-Office of Science, the gap in success rates between women and men did not close when we included all the unassigned records among men or among women.

Table 7. 5-Year Average Agency and Component Proposal Success Rates, by Sex, Including Success Rates for Records with Sex Unassigned

Agency	Unassigned[a]	Women	Women + Unassigned Records	Men	Men + Unassigned Records	Overall
Department of Agriculture-National Institute of Food and Agriculture	25%	22%	23%	21%	21%	21%
Department of Defense (DOD)-Army Research Office	43%	48%	44%	49%	47%	47%
DOD-Defense Threat Reduction Agency	14%	17%	15%	17%	16%	16%
DOD-Office of Naval Research[b]	36%	42%	38%	48%	45%	44%
Department of Energy (DOE)-Advanced Research Projects Agency-Energy[b]	5%	10%	6%	7%	6%	6%
DOE-Energy Efficiency and Renewable Energy[b]	4%	7%	5%	5%	5%	5%
DOE-Office of Science[c]	25%	34%	28%	41%	36%	36%
National Institutes of Health (NIH)[c,d]		25%		25%		
National Science Foundation (NSF)[c,d]		26%		25%		

Source: GAO analysis of agency data. | GAO-16-14

[a]"Unassigned" are those records for which the principal investigator could not be assigned a gender based on our gender-name matching algorithm.

[b]The Office of Naval Research data from 2009 were not sufficiently reliable for inclusion in this analysis; data prior to 2010 are not available for the Advanced Research Projects Agency-Energy; data prior to 2011 are not available for the Office of Energy Efficiency and Renewable Energy.

[c]This agency gives out many different categories of awards; this represents the overall success rate for all categories combined.

[d]NIH and NSF provided summary statistics based on self-reported sex averaged across all five years for which we collected data; we did not carry out a gender name match for these two agencies and therefore do not have 'unassigned' cases.

In addition, we calculated summary statistics for men, women, and records that we could not assign to a specific gender. NIH and NSF provided us with the corresponding summary statistics from their own data, including detailed descriptions of how the calculations were done so that we could ensure that our analyses were similar. In general, we found that, at all but one agency, the records that had no reported sex or for which sex could not be assigned had a lower success rate than records with an associated sex.

Analysis of Factors That May Explain Any Differences in Success Rates

In order to explore what factors might explain any differences in success rates between women and men, we relied on NSF's 2013 SDR. Using the SDR, we first calculated the observed odds ratio for women in order to compare the odds of funding for women vs. men overall. Next we controlled for the other factors that influence the receipt of federal funding (such as discipline of PhD, being employed at an educational institution, academic position, etc.) using regression analysis.

Logistic regression is the appropriate technique when the dependent variable is binary or has two categories; in this case, the two categories are: (1) individual receives federal funding, or (2) does not receive federal funding. We coded researcher's sex as a binary variable that was equal to 1 if the respondent reported being a woman and zero if the respondent reported being a man. For binary—also known as "dummy"—variables, a statistically significant odds ratio that is greater/less than 1.00 indicates that individuals with that characteristic are more/less likely to be awarded than individuals without it. That is, if women have a statistically significant odds ratio of less than 1.00, it indicates that their odds of funding are lower than men. Our fully adjusted model took into account the following characteristics:

- Discipline of PhD
- Whether one is employed full time
- Whether one is employed by an educational institution
- Whether one is a post-doc, teaching faculty, or research faculty
- Whether one works outside the field of their PhD
- Whether there are children under 6 in the household
- Whether there are children between 6-11 years old in the household

- Age
- U.S. citizenship status
- Racial or ethnic minority status

We also tested several interaction terms of sex with other characteristics to see if having children or being older, for example, had a differential impact for women than for men. We did not find any of these interaction terms to be statistically significant, and they did not improve the performance of the model. Detailed results of the model are available in appendix II.

OBJECTIVE 2 METHODS

To determine the extent to which federal agencies enforce Title IX at universities they fund for STEM research, we reviewed relevant federal laws, regulations, and guidance, interviewed agency officials, and analyzed relevant documents. We evaluated agencies' Title IX compliance activities against relevant statutory and regulatory requirements in addition to federal internal control standards for effective communication with external stakeholders.[17] Specifically, we first reviewed Title IX regulations for each of the six grant-making agencies in our review. We also reviewed relevant regulations and guidance related to Title IX enforcement from the Department of Justice (DOJ) and the Department of Education (Education), including the Title IX common rule and DOJ's Title IX Legal Manual. To describe activities agencies take to enforce Title IX compliance at their grantee universities, we interviewed officials responsible for enforcing civil rights compliance at each agency regarding their Title IX enforcement processes, results, coordination efforts with other federal agencies, and any technical assistance that these agencies provide to universities that they fund for STEM research. We also gathered and analyzed related documentation from each agency, including agency guidance on Title IX enforcement responsibilities and examples of assurance of compliance forms federal agencies collect from the entities they fund. For those agencies that conduct compliance reviews, we gathered and analyzed examples of interview guides and data requests that agencies use to conduct compliance reviews. We also reviewed all available Title IX compliance review reports for each agency from fiscal year 2010 through 2014.

Given its designation as coordinating agency for federal Title IX compliance, we interviewed officials from the DOJ's Civil Rights Division, Federal Coordination and Compliance Section. During interviews with DOJ officials, we asked questions to improve our understanding of federal Title IX enforcement responsibilities and asked about DOJ's activities to assist federal agencies in their Title IX enforcement responsibilities. We reviewed documents and guidance DOJ provides to federal agencies related to Title IX enforcement. We also interviewed officials from Education's Office for Civil Rights, which receives and investigates the most complaints of sex discrimination filed against educational institutions. Education officials offered insights on conducting Title IX compliance reviews and complaint investigations and provided important background and context for our assessment of Title IX compliance efforts at agencies that fund STEM research.

To better understand the effect of federal Title IX compliance activities on the university grantees, we interviewed Title IX coordinators and other knowledgeable staff members at four universities that receive federal STEM research funding. We selected these universities based on (1) the amount of federal STEM research funding the university receives, (2) recent federal Title IX investigations or compliance reviews at the university, and (3) whether the university is a public or private institution. We asked university officials about their experiences with federal agencies regarding Title IX compliance activities and their insight on student awareness of their rights under Title IX.

OBJECTIVE 3 METHODS

To determine the possible actions federal agencies could take to address the representation of women in federal STEM research, we first compiled a list of possible federal actions through a literature review.[18] To identify relevant publications about women in STEM fields, we performed a literature search of a number of bibliographic databases, including ABI/Inform, Educational Research Information Center (ERIC), ProQuest Dissertations and Theses, ProQuest Research Library, ProQuest Science Journals, SciSearch, Social SciSearch, Sociological Abstracts, and Web of Science. We reviewed the following document types: scholarly/peer reviewed material, government reports, dissertations, and association/nonprofit/think tank publications published from 2005 forward. We concluded our searches in February 2015. To the resulting list of publications, we added articles NIH officials recommended and articles we identified in our own background information search. From this list, we selected those publications that contained recommendations or possible actions the federal government or universities could take to address women's representation in federal STEM research.

Table 8. Experts who Participated in GAO's Review

Names	Affiliation
Cynthia Emrich	Catalyst, Inc.
Donna Dean	NIH, retired; Association for Women in Science
Donna Ginther	University of Kansas
Gary May	Georgia Institute of Technology
Holly J. Falk-Krzesinski	Elsevier
Kelly Mack	Association of American Colleges and Universities
Kerry Ann Rockquemore	National Center for Faculty Development & Diversity
Lydia Villa-Komaroff	Cytonome/ST, LLC
Marc Goulden	University of California-Berkeley
Maria Klawe	Harvey Mudd College
Mary Ann Mason	University of California, Berkeley, School of Law
Renetta Garrison Tull	University of Maryland, Baltimore County
Ruta Sevo	NSF, retired
Scott Franklin	Rochester Institute of Technology
Shirley Malcom	American Association for the Advancement of Science
Stephen Ceci	Cornell University
Vivian Pinn	NIH, retired
Wendy Baldwin	NIH, retired
Wendy M. Williams	Cornell University

Source: GAO. | GAO-16-14

We then asked 25 subject-matter experts to validate this list of possible federal agency actions using a data collection instrument we pre-tested with a university-based STEM researcher outside of GAO who was familiar with our review.[19] We compiled our initial list of experts by asking for recommendations during our early interviews and noting authors of seminal studies and reports and members of relevant panel discussions or workshop events. We included individuals representing universities, nonprofits conducting research on women's issues, academic research associations, and others in the scientific community. We also included retired federal officials from agencies we reviewed who are familiar with federal grant programs. We sought gender balance to the extent possible. Nineteen experts responded to our request and completed our data collection instrument. This report provides our analysis of the overall evidence gathered through our expert consultations. The report findings do not necessarily represent the views of specific individuals or institutions. See table 8 for a list of the experts who participated in our review.

Based on feedback from experts, we narrowed our initial list from 8 categories of 29 possible actions to 4 categories of 13 possible actions. We arrived at this final list by combining similar categories and actions and in some cases, moving actions between their original categories. We narrowed our list based on feedback from experts, including combining overlapping actions, removing actions that experts felt would not achieve their intended goals, and revising the text to improve clarity. We added one action to this list subsequent to our expert consultations: conducting blind reviews of research proposals.[20] We sent the final list to federal officials from each grant-making agency in our review and conducted interviews with agency officials to discuss the actions their agencies had taken. Following these interviews, we asked officials to provide written input on which actions their agencies had taken. For those actions they had not taken, we asked them to provide information on the feasibility of implementing the actions in the future. We did not evaluate the efficacy of these actions and this report does not establish whether they will achieve their intended goals or whether they would be suited to each agency's particular organizational context.

Finally, to supplement our research, we obtained perspectives on the federal STEM research grant application process by conducting a group interview with five female STEM researchers. To identify participants for our group interview, we asked the Society of Women Engineers (SWE) to identify volunteers to participate. The volunteers were from SWE's Women in Academia Committee or in the networks of women in that committee. In addition, we included in this group interview one American Association for the Advancement of Science fellow with an Applied Physics background who was suggested by an official from the Department of Education.

APPENDIX II: ADDITIONAL ANALYSIS OF NATIONAL SCIENCE FOUNDATION'S SURVEY OF DOCTORATE RECIPIENTS

We analyzed data from the National Science Foundation's (NSF) 2013 Survey of Doctorate Recipients (SDR) to estimate the difference in the likelihood of receiving funding for their work from any federal agency, and specifically from the National Institutes of Health (NIH), NSF, the Department of Defense (DOD), and the Department of Energy (DOE), between women and men working in STEM fields.[1] We first estimated the unadjusted sex differences in receiving federal funding by examining the numbers and percentages of women and men that received funding, and odds and odds ratios derived from them, as shown in table 9 below.[2] However, as demonstrated in figure 10, women and men are represented differently among doctorate recipients in the various STEM disciplines. Therefore, we re-estimated the differences between women and men in the likelihood of receiving funding from the four agencies in our review captured by the SDR using multivariate models which controlled for the doctorate recipient's academic discipline. In additional multivariate models, we also controlled for a variety of other factors, including age, citizenship status, minority status, the type of position they occupied, and whether the individual had a post-doctoral fellowship, worked full-time, or had children in the household. Because the outcome variable is dichotomous, we used a logistic regression model which we estimated using maximum likelihood. These regression results are shown in tables 10 through 13.

Unadjusted (Bivariate) Differences

Table 9 indicates that, in 2013, roughly 231,000 of the 838,000 U.S trained STEM doctorates, or about 27.5 percent of them, were receiving some sort of federal funding. Roughly 10 percent of STEM doctorates received funding from NIH, slightly more than 6 percent from NSF, slightly fewer than 6 percent from DOD, and about 3 percent received funding from DOE.

Differences in the likelihood of receiving federal funding between women and men from any federal agency were statistically significant, but not pronounced. The top section of table 9 shows that 28.2 percent of men versus 26.2 percent of women doctorates received funding from at least one federal agency. An alternative way of expressing this difference is by calculating the odds of receiving funding for women and men, which indicates how many women and men received funding for every woman and man that did not. The odds of receiving funding from any federal agency for women were 72,211 ÷ 203,328 = 0.36, indicating that .36 women received funding for every one woman that did not. By comparison, the odds of receiving funding from any federal agency for men were 158,561 ÷ 403,839 = 0.39. The ratio of these two odds, or the odds ratio, which is 0.36 ÷ 0.39 = 0.91, apart from rounding, indicates that women were less likely than men to receive any federal funding, by a factor of 0.91. An odds ratio of 1.0 would indicate no difference between women and men.

Table 9. Differences Between Women and Men Who Received Federal Funding for STEM Research, and Funding from NIH, NSF, DOD and DOE

Sex	No Federal Funding	Federal Funding	Total	Odds of Funding	Odds Ratio – Women: Men[a,b]
Men	403,839	158,561	562,400	0.393	
	71.8%	28.2%	100.0%		
Women	203,328	72,211	275,539	0.355	0.905***
	73.8%	26.2%	100.0%		
Total	607,167	230,773	837,939		
	72.5%	27.5%	100.0%		
Sex	No NIH Funding	NIH Funding	Total	Odds of Funding	Odds Ratio – Women: Men
Men	512,185	50,216	562,400	0.098	
	91.1%	8.9%	100.0%		
Women	243,473	32,066	275,539	0.132	1.343***
	88.4%	11.6%	100.0%		
Total	755,657	82,282	837,939		
	90.2%	9.8%	100.0%		
Sex	No NSF Funding	NSF Funding	Total	Odds of Funding	Odds Ratio – Women: Men
Men	523,300	39,100	562,400	0.075	
	93.1%	7.0%	100.0%		
Women	261,383	14,156	275,539	0.054	0.725***
	94.9%	5.1%	100.0%		
Total	784,683	53,256	837,939		
	93.6%	6.4%	100.0%		
Sex	No DOD Funding	DOD Funding	Total	Odds of Funding	Odds Ratio – Women: Men
Men	523,751	38,649	562,400	0.074	
	93.1%	6.9%	100.0%		
Women	265,538	10,001	275,539	0.038	0.510***
	96.4%	3.6%	100.0%		
Total	789,289	48,650	837,939		
	94.2%	5.8%	100.0%		
Sex	No DOE Funding	DOE Funding	Total	Odds of Funding	Odds Ratio – Women: Men
Men	539,485	22,916	562,400	0.042	
	95.9%	4.1%	100.0%		
Women	271,665	3,874	275,539	0.014	0.336***

Table 9. (Continued)

Sex	No Federal Funding	Federal Funding	Total	Odds of Funding	Odds Ratio – Women: Men[a,b]
	98.6%	1.4%	100.0%		
Total	811,149	26,790	837,939		
	96.8%	3.2%	100.0%		

Source: GAO analysis of 2013 SDR data. | GAO-16-14

Notes: Numbers in a given row may not add up to 100% due to rounding.

[a]Asterisks denote the level of statistical significance; i.e., *** p<0.01, ** p<0.05, * p<0.1.

[b]The female:male odds ratio indicates the difference between women and men in the odds of receiving funding. For example, a female:male odds ratio of .75 would indicate that women have lower odds of receiving funding than men, by a factor of 0.75, or that female STEM PhDs have a 25% lower chance than men of receiving funding, on average, from this particular agency. The ratios do not reflect any adjustments for qualifications or other researcher characteristics.

There were, however, significant and pronounced differences in each of the four specific agencies we considered. Female STEM doctorate holders were more likely than male doctorates to receive funding from NIH, by a factor (odds ratio) of 1.34, but less likely to receive funding from NSF, DOD, and DOE, by factors of 0.73, 0.51, and 0.34, respectively. At the latter two agencies the differences were especially pronounced. The odds ratios reflecting the differences imply that female doctorates were only roughly half as likely as their male counterparts to receive funding from DOD, and that female doctorate holders were only one-third as likely as their male counterparts to receive funding from DOE.

Differences in Academic Discipline between Women and Men

One advantage of using odds ratios to estimate the differences between women and men is that they can be adjusted, or re-estimated, using multivariate models which simultaneously control for differences between men and women in other characteristics which affect the likelihood of receiving funding. The multivariate analyses we report below show that the discipline in which doctorates were earned is very strongly—albeit differently–related to whether doctorate holders receive funds from each of the four specific agencies. Because of that, we conclude that the differences in the disciplines in which doctorates are earned by women and men is strongly related to the differences between women and men in receiving federal funds from these agencies. Figure 10 below shows that women and men earn doctorates in the various STEM disciplines in different proportions. Figure 10 shows that 44 percent of male PhDs received degrees in engineering or the physical sciences, while only 21 percent of male PhDs received degrees in the social sciences. By comparison, only 18 percent of female PhDs received degrees in engineering or the physical sciences, while nearly 40 percent of female PhDs received degrees in the social sciences.

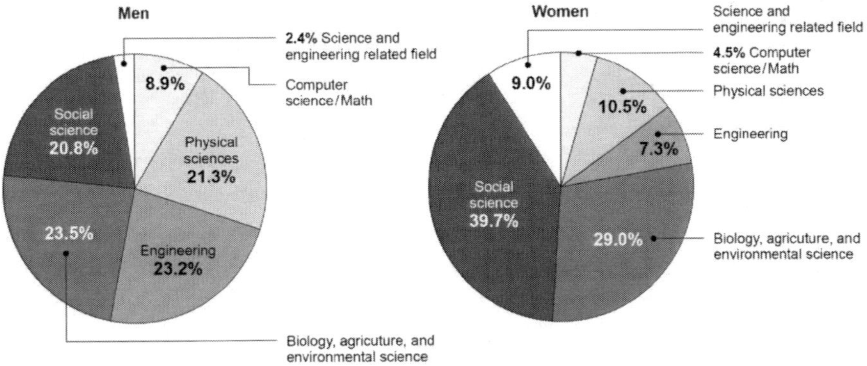

Source: GAO analysis of 2013 Survey of Doctorate Recipients (SDR) data. | GAO-16-14.

Figure 10. Male and Female PhDs in STEM Fields, by Discipline.

Adjusted Differences at the Four Agencies

In summary, after the full set of controls, there is no statistically significant difference between men and women in the chances of receiving funding from NIH, a significant but modest difference at NSF, where women's chances of receiving funding were 12 percent lower than men's, and significant and more sizable differences at DOD and DOE that indicate women's chances of receiving funding were roughly 23 percent lower than men's at DOD and roughly 38 percent lower at DOE. Figure 11 summarizes the results of estimating the differences between women and men in the likelihood of receiving funding from these four federal agencies, before and after adjusting for differences in the academic disciplines and other characteristics of female and male doctorate holders. The unadjusted differences between women and men vary greatly across the four agencies.

Women have a substantially and significantly higher likelihood of receiving funding than men at NIH, but substantially and significantly lower likelihoods of receiving funding than men at the other three agencies. When differences in academic disciplines between women and men are taken into account, the differences in their likelihoods of receiving funding from each of the four agencies decrease. In the model that adjusts for discipline and the other factors, the difference between women and men is rendered insignificant at NIH, where women have a 4 percent lower chance of receiving funding. At NSF, controlling for these other factors results in women's chances of receiving funding declining relative to men.[3] At DOD and DOE, these latter adjustments have little effect, and the difference between women and men receiving funding remain much as they were when only discipline was controlled.

In tables 10 through 13 below, we show estimates of adjusted as well unadjusted differences between women and men in the likelihoods of receiving funding from NIH (table 10), NSF (table 11), DOD (table 12) and DOE (table 13), as well as estimates of the effects of the other factors, or characteristics, of the doctorate recipients that have been adjusted for. In all cases the estimated coefficients are odds ratios from logistic regression models which have been fitted to the weighted SDR data using maximum likelihood estimation. Model (1) in each table shows the unadjusted odds ratio estimating the difference between women and men from a simple bivariate model which regresses the odds of receiving funding from the different agencies on sex alone. This is in each case identical to the odds ratio that was produced from the observed data in table 9 (above). Model (2) in each table regresses the odds of receiving funding from the different agencies on sex and academic discipline, where disciplines are represented by dummy variables that contrast 1) computer science and mathematics, 2) biology, agriculture and environmental sciences, 3) physical sciences, 4) engineering, 5) science and engineering related fields, and 6) non-science and engineering related fields with 7) social science, which is the omitted or reference category. Finally, model (3) in each table regresses the odds of receiving funding from the different agencies on sex, academic discipline, and all of the other factors we controlled for (i.e., age, citizenship status, minority status, full- or part-time work schedules, employment by an educational institution, occupying a teaching faculty position, holding a post-doctoral fellowship, and having children in the household). Below we present the results of each model, by agency.

NIH

Although women have significantly higher unadjusted odds of receiving funding from NIH, when adjusted for educational and other characteristics, the difference between women and men in odds of receiving funding disappears. Model (1) in table 10 shows – as we saw in table 9 – that women were more likely than men to receive funding from NIH, by a factor of 1.34, at least when sex is considered by itself and other characteristics are ignored. Model (2) shows that there were large differences between STEM doctorate holders from different disciplines in the likelihood of receiving funding from NIH. For example, those with doctorates in computer science/mathematics disciplines and in engineering were less likely than doctorates in the social sciences to receive funding from NIH, by factors of 0.73 and 0.74, respectively, while those with doctorates in biology, agriculture and environment disciplines were more likely than social scientists to receive funding from NIH, by a factor of 4.35. Importantly, when discipline is taken into account, the odds ratio associated with the difference between women and men in receiving funding diminishes substantially, from 1.34 to 1.10, though it remains significant. Model (3) includes a number of other factors, many of which have sizable and significant effects on receiving funding from NIH, such as being employed by an educational institution (odds ratio = 5.29), having been a postdoctoral fellow (odds ratio = 2.47), and having a teaching faculty position (odds ratio = 0.27). When these effects are taken into account the difference between women and men of receiving funding is reduced from 1.10 to 0.96, and is no longer statistically significant.

NSF

When educational and other characteristics are taken into account at NSF, the significant differences in the odds of receiving funding between women and men diminishes, but does not entirely disappear. Model (1) in table 11 shows that women were less likely than men to receive funding from NSF, by a factor of 0.73, at least when other characteristics are ignored. Here, too, Model (2) shows large differences between doctorate holders from different disciplines in the likelihood of receiving funding from NSF; for example, those with doctorates in computer science/mathematics disciplines, in the physical sciences, and in engineering were more likely than social scientists to receive funding from NSF, by factors of 4.33, 3.82, and 2.77, respectively. When discipline differences are taken into account at NSF the odds ratio associated with the difference between women and men in receiving funding diminishes substantially, from 0.73 to 1.02, and becomes statistically insignificant.[4] As at NIH, many of the other factors included in Model (3) have sizable and significant effects on receiving funding from NSF, including being employed by an educational institution (odds ratio = 15.78), and working outside of their field (odds ratio = 0.37). When these effects are taken into account the difference between women and men receiving funding from NSF returns to being statistically significant (odds ratio = 0.88), and indicates that women have modestly lower chance than men of receiving funding.

DOD

Controlling for discipline of PhD and other characteristics decreases the gap between women and men in likelihood of receiving funding from DOD, though a statistically significant difference of about 23 percent in the chance of receiving funding remains in the fully adjusted model. Model (1) in table 12 shows that women were less likely than men to receive funding from DOD, by a factor of 0.51, when other characteristics are ignored. Model (2) shows large differences between doctorates from different disciplines in the likelihood of receiving funding from DOD; doctorate holders in computer science/mathematics disciplines, in the physical sciences, and in engineering were more likely than social scientists to receive funding from DOD, by factors of 3.59, 3.16, and 5.44, respectively. When discipline differences are taken into account at DOD the odds ratio associated with the difference between women and men in receiving funding from DOD diminishes from 0.51 to 0.78, but remains significant. At DOD, as elsewhere, many of the other factors included in Model (3) have sizable and significant effects on receiving funding. For example, at DOD being a teaching faculty member diminishes the likelihood of receiving funding (odds ratio = 0.50), and being a citizen doubles the odds of receiving funding. When these effects are taken into account the difference between women and men in the odds of receiving funding from DOD is about 23 percent (odds ratio =.772) and remains statistically significant.

DOE

When educational and other characteristics are taken into account, about half of the difference between women and men in the odds of receiving funding from DOE is explained, though a statistically significant gap remains in the fully adjusted model. Model (1) in table 13 shows that women were less likely than men to receive funding from DOE, by a factor of 0.34 when other characteristics are ignored. Model (2) shows large differences between doctorates from different disciplines in the likelihood of receiving funding from DOE; those with doctorates in computer science/mathematics disciplines, in the physical sciences, and in engineering were more likely than those with doctorates in the social sciences to receive funding from DOE, by factors of 4.04, 15.0, and 14.92, respectively. When discipline differences are taken into account at DOE the odds ratio associated with the difference between women and men in receiving funding diminishes from 0.34 to 0.62, but remains significant. At DOE, as elsewhere, many of the other factors included in Model (3) have sizable and significant effects on receiving funding. Working full time, for example, increases the likelihood of receiving DOE funding (odds ratio = 3.50), while being on a teaching faculty diminishes the likelihood of receiving funding (odds ratio = 0.55). When these effects are taken into account, the difference between women and men in the odds of receiving funding does not change in size and statistical significance.

Limitations

While elsewhere in this report we have focused on differences between men and women in their chances of receiving federal grants, the SDR data do not distinguish grants and contracts, and allow us only to examine gender differences in receiving federal funding from grants and contracts combined. To the extent that gender differences in the likelihood of receiving grants are unlike gender differences in receiving contracts, the results here may not be comparable to results reported in other parts of this report. Moreover, the data are self-reported, and thus prone to error resulting from inaccurate survey responses. Respondents were first asked whether any of their work during 2012 was supported by contracts or grants from the U.S. government, and those who said yes were then asked which federal agencies supported their work. Respondents might have mistakenly "telescoped" support from periods prior to 2012, mistakenly reported where the support came from, or not recalled support correctly in other ways. Our analyses are inherently limited as well, since in estimating gender differences in funding and attempting to control for potentially confounding factors, we lack complete information on such confounds. Most notably, the SDR data do not allow us, in looking at the likelihood of receiving federal funding, to restrict our attention to persons who apply. Thus, the differences between men and women in the likelihood of receiving federal funding shown here may result from differences between men and women in the likelihood of applying and any number of other omitted variables.

Table 10. Odds Ratios Estimating the Effects of Sex and Other Factors on Receiving NIH Funding

	Odds Ratios - NIH Funding		
Variables	**Model (1)**	**Model (2)**	**Model (3)**
Female	1.343***	1.095**	0.964
	(0.0546)	(0.0480)	(0.0459)
PhD in Computer Science/Mathematics		0.725***	0.594***
		(0.0871)	(0.0740)
PhD in Biology, Agriculture and Environmental Science		4.348***	3.542***
		(0.245)	(0.213)
PhD in Physical Sciences		0.927	0.873
		(0.0745)	(0.0744)
PhD in Engineering		0.744***	0.747***
		(0.0642)	(0.0675)
PhD in Social Science		REF	REF
PhD in Science and Engineering Related Field		3.219***	2.970***
		(0.278)	(0.277)
PhD in non-Science and Engineering Related Field		2.864	3.432
		(2.443)	(2.777)
Working full time			1.946***
			(0.149)
Employed by Educational Institution			5.285***
			(0.274)
Post Doc			2.471***
			(0.219)
Teaching Faculty			0.270***
			(0.0199)
Works outside of field			0.454***
			(0.0557)
Children under 6 in household			1.035
			(0.0631)
Children 6-11 in Household			1.029
			(0.0614)
Age			0.980***
			(0.00238)
Citizenship status			1.160**
			(0.0830)
Minority			0.798***

	Odds Ratios - NIH Funding		
Variables	Model (1)	Model (2)	Model (3)
			(0.0493)
Constant	0.0980***	0.0602***	0.0460***
	(0.00254)	(0.00317)	(0.00817)
Sample Total	30,696	30,696	30,696
Estimated Population Total	837,939	837,939	837,939
F	52.70	202.2	184.6

Source: GAO analysis of 2013 SDR data. | GAO-16-14

Notes: Standard errors are in parentheses. Asterisks denote the level of statistical significance; i.e., *** p<0.01, ** p<0.05, * p<0.1. The referent category for the polytomous variable indicating discipline (or PhD type) is Social Science. Except for age (in years), which was coded linearly, all other variables are dichotomous, and the referent category is the remaining category not described by the labeled category shown in the table (e.g., male, not working full time, not employed by an educational institution, etc.).

Table 11. Odds Ratios Estimating the Effects of Sex and Other Factors on Receiving NSF Funding

		Odds Ratios - NSF Funding	
Variables	Model (1)	Model (2)	Model (3)
Female	0.725***	1.024	0.878**
	(0.0384)	(0.0582)	(0.0532)
PhD in Computer Science/Mathematics		4.328***	4.154***
		(0.427)	(0.427)
PhD in Biology, Agriculture and Environmental Science		1.933***	1.709***
		(0.165)	(0.153)
PhD in Physical Sciences		3.822***	5.337***
		(0.324)	(0.475)
PhD in Engineering		2.776***	4.402***
		(0.250)	(0.423)
PhD in Social Science		REF	REF
PhD in Science and Engineering Related Field		0.300***	0.248***
		(0.0867)	(0.0727)
PhD in non-Science and Engineering Related Field		-	-

Table 11. (Continued)

		Odds Ratios - NSF Funding	
Variables	**Model (1)**	**Model (2)**	**Model (3)**
Working full time			2.318***
			(0.248)
Employed by Educational Institution			15.78***
			(1.275)
Post Doc			0.680***
			(0.0745)
Teaching Faculty			0.617***
			(0.0402)
Works outside of field			0.372***
			(0.0701)
Children under 6 in household			1.076
			(0.0759)
Children 6-11 in household			1.169**
			(0.0779)
Age			0.986***
			(0.00281)
Citizenship status			1.023
			(0.0806)
Minority			0.826**
			(0.0633)
Constant	0.0747***	0.0302***	0.00506***
	(0.00219)	(0.00229)	(0.00113)
Sample Total	**30,696**	**30,684**	**30,684**
Estimated Population Total	837,939	837,640	837,640
F	36.95	68.99	101.7

Source: GAO analysis of 2013 SDR data. | GAO-16-14

Notes: Standard errors are in parentheses. Asterisks denote the level of statistical significance; i.e., *** p<0.01, ** p<0.05, * p<0.1.The referent category for the polytomous variable indicating discipline (or PhD type) is Social Science. Additionally, as there were zero cases of individuals with non-Science and Engineering related doctorates indicating they had received funding from NSF, this category was automatically dropped from the specification. Except for age (in years), which was coded linearly, all other variables are dichotomous, and the referent category is the remaining category not described by the labeled category shown in the table (e.g., male, not working full time, not employed by an educational institution, etc.).

Table 12. Odds Ratios Estimating the Effects of Sex and Other Factors on Receiving DOD Funding

	Odds Ratios - DOD Funding		
Variables	**Model (1)**	**Model (2)**	**Model (3)**
Female	0.510***	0.781***	0.772***
	(0.0311)	(0.0504)	(0.0515)
PhD in Computer Science/Mathematics		3.594***	3.555***
		(0.408)	(0.408)
PhD in Biology, Agriculture and Environmental Science		1.665***	1.430***
		(0.166)	(0.144)
PhD in Physical Sciences		3.162***	3.014***
		(0.306)	(0.297)
PhD in Engineering		5.440***	5.157***
		(0.500)	(0.500)
PhD in Social Science		REF	REF
PhD in Science and Engineering Related Field		1.047	0.960
		(0.201)	(0.185)
PhD in non-Science and Engineering Related Field		4.713	4.508
		(4.950)	(4.835)
Working full time			2.226***
			(0.209)
Employed by Educational Institution			1.231***
			(0.0814)
Post Doc			1.116
			(0.173)
Teaching Faculty			0.499***
			(0.0505)
Works outside of field			0.785**
			(0.0844)
Children under 6 in household			0.890
			(0.0693)
Children 6-11 in household			0.988
			(0.0711)
Age			0.992***
			(0.00284)

Table 12. (Continued)

	Odds Ratios - DOD Funding		
Variables	**Model (1)**	**Model (2)**	**Model (3)**
Citizenship status			2.095***
			(0.199)
Minority			0.778***
			(0.0662)
Constant	0.0738***	0.0262***	0.0122***
	(0.00218)	(0.00219)	(0.00278)
Sample Total	**30,696**	**30,696**	**30,696**
Estimated Population Total	837,939	837,939	837,939
F	121.5	83.25	49.07

Source: GAO analysis of 2013 SDR data. | GAO-16-14

Notes: Standard errors are in parentheses. Asterisks denote the level of statistical significance; i.e., *** p<0.01, ** p<0.05, * p<0.1. The referent category for the polytomous variable indicating discipline (or PhD type) is Social Science. Except for age (in years), which was coded linearly, all other variables are dichotomous, and the referent category is the remaining category not described by the labeled category shown in the table (e.g., male, not working full time, not employed by an educational institution, etc.).

Table 13. Odds Ratios Estimating the Effects of Sex and Other Factors on Receiving DOE Funding

	Odds Ratios - DOE Funding		
Variables	**Model (1)**	**Model (2)**	**Model (3)**
Female	0.336***	0.620***	0.622***
	(0.0307)	(0.0594)	(0.0612)
PhD in Computer Science/Mathematics		4.040***	3.864***
		(0.959)	(0.916)
PhD in Biology, Agriculture and Environmental Science		3.024***	2.608***
		(0.606)	(0.526)
PhD in Physical Sciences		15.00***	15.32***
		(2.755)	(2.835)
PhD in Engineering		14.92***	15.14***
		(2.761)	(2.850)
PhD in Social Science		REF	REF

	Odds Ratios - DOE Funding		
Variables	**Model (1)**	**Model (2)**	**Model (3)**
PhD in Science and Engineering Related Field		1.748	1.530
		(0.649)	(0.569)
PhD in non-Science and Engineering Related Field		24.78***	23.56***
		(26.21)	(26.49)
Working full time			3.495***
			(0.495)
Employed by Educational Institution			1.746***
			(0.148)
Post Doc			1.006
			(0.191)
Teaching Faculty			0.546***
			(0.0676)
Works outside of field			0.605***
			(0.0971)
Children under 6 in household			1.048
			(0.106)
Children 6-11 in household			0.901
			(0.0873)
Age			1.001
			(0.00383)
Citizenship status			1.508***
			(0.173)
Minority			0.782**
			(0.0949)
Constant	0.0425***	0.00547***	0.00116***
	(0.00161)	(0.000977)	(0.000383)
Sample Total	30,696	30,696	30,696
Estimated Population Total	837,939	837,939	837,939
F	142.0	80.19	43.64

Source: GAO analysis of 2013 SDR data. | GAO-16-14

Notes: Standard errors are in parentheses. Asterisks denote the level of statistical significance; i.e., *** $p<0.01$, ** $p<0.05$, * $p<0.1$. The referent category for the polytomous variable indicating discipline (or PhD type) is Social Science. Except for age (in years), which was coded linearly, all other variables are dichotomous, and the referent category is the remaining category not described by the labeled category shown in the table (e.g., male, not working full time, not employed by an educational institution, etc.).

Figure 11. Odds Ratios Estimating Sex Difference in the Likelihood of Receiving Federal Funding for 3 Separate Models.

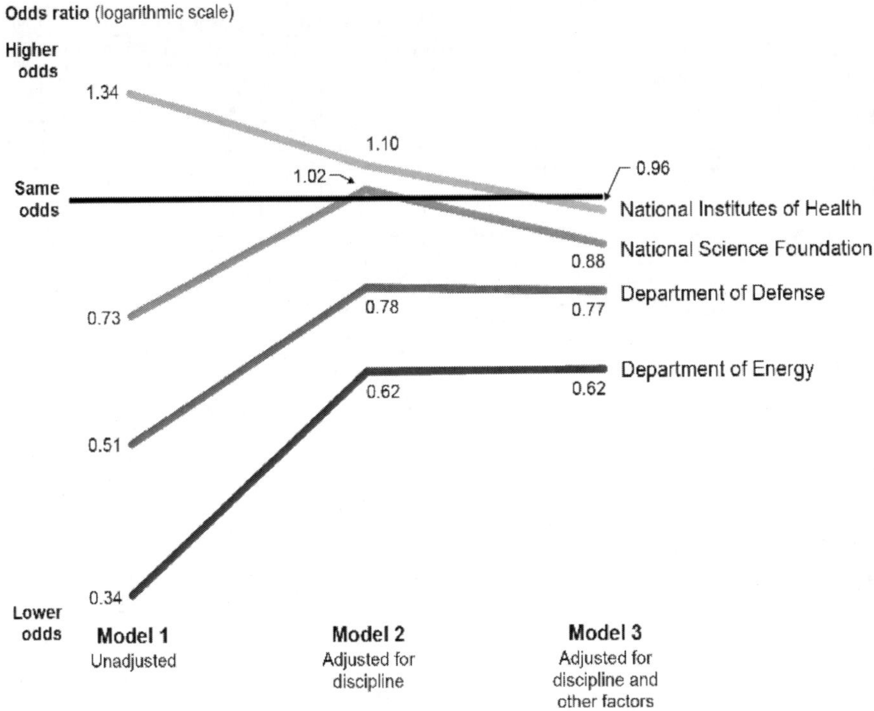

Source: GAO analysis of 2013 Survey of Doctorate Recipients (SDR) data. | GAO-16-14.

APPENDIX III: SUMMARIES OF SELECTED AGENCY RESPONSES ON GAO TABLE OF ACTIONS

The following tables include the actions that we identified based on a literature review, meetings with leading agencies, and consulting with experts. We asked the six selected agencies to provide information on the actions they had taken and, for those actions they had not taken, agencies provided information on the feasibility of taking such actions in the future.[1]

Table 14. Possible Actions We Identified Through a Literature Review and Expert Consultation Related to Agency Leadership and Collaboration for Federal STEM Grant-making Agencies

Action	Description	Summaries of Selected Agency Responses to Identified List of Actions
Develop institutional leadership that values diversity	Develop institutional leadership that values diversity, in part through policies requiring existing and potential employees to demonstrate cultural competence and a commitment to diversity.	• Department of Agriculture-National Institute for Food and Agriculture (USDA-NIFA) reported that its senior leadership attends diversity and cultural competency training to enhance awareness and to increase emotional intelligence. Additionally, NIFA reported that USDA developed an agency-wide policy that requires participation in cultural transformation initiatives to be included in each employee's performance plan, and that USDA staff, including senior leadership, are rated annually by their supervisors on the extent to which they support the agency's diversity initiatives. • Department of Defense (DOD) Personnel and Readiness (P&R) reported that DOD published the Department's Diversity and Inclusion Strategic Plan 2012-2017 and that DOD has promoted specific programs that target cultural competency and commitment, such as the STEM Connector Million Women Mentor initiative and the Human Goals Charter, and events which target underrepresented groups. • Department of Energy (DOE) Office of Science reported that it has an agency-wide Diversity and Inclusion Strategic Plan and that the Secretary issued a Statement on Policy and Inclusion in April 2015. • National Aeronautics and Space Administration (NASA) reported that it has a Diversity Policy Statement, a Diversity and Inclusion Strategic Plan, and an agency-wide Diversity and Inclusion Strategic Partnership.

Table 14. (Continued)

Action	Description	Summaries of Selected Agency Responses to Identified List of Actions
		• National Institutes of Health (NIH) reported that it has created a strategic plan for diversity and inclusion and that the NIH Common Fund has established a program called "Enhancing the Diversity of the NIH-Funded Workforce." • National Science Foundation (NSF) reported that NSF has created a strategic plan for diversity and inclusion and that Director publishes a Policy Statement on Diversity and Inclusion. In addition, NSF's merit review criterion "Broader Impacts" seeks to foster "full participation of women, persons with disabilities, and underrepresented minorities in science, technology, engineering, and mathematics…"
Establish an agency-level working group	Establish a working group that reports directly to the agency head and has agency-wide scope to assess successful strategies and develop programs and policies that aid the recruitment, retention, and advancement of women in STEM research.	• USDA-NIFA has a Diversity and Inclusion workgroup to promote diversity awareness in the workplace and reinforce cultural competence. • DOD's Acquisition, Technology and Logistics (AT&L) reported that it does not currently have such a working group but could, if directed, alter the charter of one of its STEM or Basic Research Office working groups to include a focus on women in STEM research. DOD officials also noted that there is a DOD-wide STEM working group that may look at underrepresented groups, including women. • DOE has not taken this action and reported that if it were to take this action, it would be best taken at the program office level rather than at the agency level, given the various types of research funded at DOE.

Action	Description	Summaries of Selected Agency Responses to Identified List of Actions
		• NASA has an informal senior level group to assess the posture of the agency with regard to retention and advancement of women in STEM. • NIH has an agency-wide working group on women in biomedical careers to address barriers for women in science and develop programs and policies that aid the recruitment, retention, and advancement of women in biomedical faculty and leadership positions. NIH also reported that as of January 2015, NIH institutes, centers, and offices may include women as eligible candidates in faculty-level, diversity-targeted programs to address faculty recruitment, appointment, retention, or advancement. • NSF reported that the NSF Broadening Participation Working Group is charged to examine programmatic/strategic challenges and opportunities as well as to recommend strategic actions to lead the nation forward in developing a globally engaged, diverse workforce.
Maintain publicly-available websites on women in STEM research	Maintain a publicly-available website that contains information for stakeholders on supporting women in STEM research careers, including information for researchers, on mentoring and career development opportunities.	• USDA-NIFA reported that it participates in interagency groups promoting STEM education that are developing publicly-available websites that contain information for stakeholders on supporting women in STEM. • DOD does not have such a website, but reported it could make this information available on its website, if necessary. • DOE Office of Science reported that the Office of Economic Impact & Diversity hosts a website that features biographical information about women in the DOE system. • NASA has a Women@NASA site, a One Stop Shopping Initiative site, and the Student Corner of its MissionSTEM site.

Table 14. (Continued)

Action	Description	Summaries of Selected Agency Responses to Identified List of Actions
		• NIH developed the Women of Color Research Network (WoCRN), a new social media site providing information, mentoring, and career development opportunities for women of color in biomedical careers and for all supporters of diversity in the scientific workforce. • NSF supports several web-based portals. For example, the NSF Career Life Balance Initiative website was designed to communicate NSF's efforts to clear the obstacles from the STEM Career-Life pathways leading from graduate education through full professor.
Establish and highlight best practices	Partner with institutional stakeholders, such as professional organizations and university grantees, to establish and highlight best practices for flexible work arrangements.	• While USDA-NIFA does not currently collect best practices, it noted it could partner with land-grant institutions at its annual conference to identify best practices. • DOD P&R reported that DOD partners with other agencies such as Office of Personnel Management, Office of Science and Technology of the President, Department of Labor, Department of Justice and others, to effectively implement strategies for flexible work arrangements and improve awareness. • DOE Office of Science reported that DOE has partnered with its stakeholders to highlight strategies via workshops. • NASA partners with grant recipients on reporting best practices, including family-friendly policies for researchers, among grantees through the MissionSTEM website. • NIH recently partnered with the Food and Drug Administration during National Women's Health Week 2015 to host a panel event to recognize the importance of women and diversity in clinical research.

Action	Description	Summaries of Selected Agency Responses to Identified List of Actions
		• NSF reported that it is a member of a public-private cooperative initiative involving 10 federal agencies and over 150 institutional organizations known as the Federal Demonstration Partnership (FDP). One of FDP's current activities is the Family Worklife Balance Project, a joint collaboration involving FDP, NIH and NSF to explore best practices, policies, and guidance on a federal and institutional level to help researchers achieve a good balance between their home life and work life.

Source: GAO analysis of literature, expert input, and federal agency interviews and data. | GAO-16-14.

Table 15. Possible Actions We Identified Through a Literature Review and Expert Consultation Related to Establishing Family-friendly Policies for Federal STEM Grantees

Action	Description	Summaries of Selected Agency Responses to Identified List of Actions
Cover certain dependent-care costs	Establish policies that allow grants to cover limited dependent care costs for researchers, including graduate students and postdocs (e.g., during off-site and after-hours meetings).	• Department of Agriculture-National Institute for Food and Agriculture (USDA-NIFA) reported that it follows USDA Departmental Regulations at 2 C.F.R. 400 and the Uniform Guidance requirements at 2 C.F.R. 200 incorporated by reference in the award terms and conditions to determine allowability of dependent care costs. • Department of Defense's (DOD) Acquisition, Technology and Logistics (AT&L) reported that the Office of Management and Budget (OMB) cost principles in 2 C.F.R. part 200, Subpart E, contain certain allowances for dependent-care costs. DOD AT&L also noted that the costs would need to be consistent with the grantee institution's own policies in order to be chargeable to federal awards.

Table 15. (Continued)

Action	Description	Summaries of Selected Agency Responses to Identified List of Actions
		• Department of Energy (DOE) Office of Science reported that DOE follows 2 C.F.R. 200 (Uniform Administrative Requirements, Cost Principles, and Audit Requirements for Federal Awards), which contains several family friendly provisions. The DOE Office of Management (MA) served on the interagency groups that developed 2 C.F.R. 200, a regulation issued December 26, 2014. The MA office also participated in the development by the Research Business Models Working Group of the Committee on Science of the National Science and Technology Council of soon-to-be-published Frequently Asked Questions about family friendly policies that apply across agencies. • National Aeronautics and Space Administration (NASA) reported that it does not currently have a supplemental policy in place and that cost allowability under grants is governed by Uniform Administrative Requirements, Cost Principles, and Audit Requirements for Federal Awards (2 C.F.R. Part 200). These costs may be allowable where the grantee offers such as part of their fringe benefits, as well as the special cases set forth in 2 C.F.R. §§200.432 and 200.473. However, NASA does not have special authority to allow additional costs of this type, so is limited to paying what is generally allowable under 2 C.F.R. Part 200. • National Institutes of Health (NIH) grant awards (with the exception of National Research Service Awards) allow for reimbursement of actual, allowable costs incurred for child care provided such costs are incurred under formally-established institutional policies that are consistently applied regardless of the source of support.

Action	Description	Summaries of Selected Agency Responses to Identified List of Actions
		NIH is also implementing a new policy allowing institutional grantees to use part of the grant funding to help cover daycare costs and requires that applications for conference support must include descriptions of child and family care offerings at the conference site. • National Science Foundation (NSF) reported that OMB is responsible for issuance of cost policies for assistance awards. Updated requirements were implemented by OMB on December 26, 2014, in the new Uniform Administrative Requirements, Cost Principles, and Audit Requirements for Federal Awards (2 C.F.R. § 200). The Uniform Guidance addresses dependent care costs in 2 C.F.R. § 200.432, Conferences, and 2 C.F.R. § 200.473, Transportation Costs. NSF's coverage regarding dependent care costs is contained in the NSF Proposal & Award Policies & Procedures Guide and is consistent with that established in the Uniform Guidance.
Provide support for transitions back to career after family leave	Provide grants or supplements for researchers to transition back to their careers after returning from family leave. For example, these awards could provide support to update research skills or participate in a customized mentoring program.	• USDA-NIFA officials reported that it has no grants or supplements for this purpose. • DOD AT&L reported that DOD grant funding must be directly related to the agency's mission, which would preclude it from providing this type of support. • DOE did not report having funding specifically for this purpose but noted that the institution can apply for a regular research grant or a supplement to an existing research grant. • NASA reported that this support could only be provided if it is part of the authorizing legislation and fits into NASA's mission. NASA also noted that its Research Opportunities in Space and Earth Sciences (ROSES) solicitation accommodates all reasonable requests to re-phase ROSES awards (e.g., for family or medical leave).

Table 15. (Continued)

Action	Description	Summaries of Selected Agency Responses to Identified List of Actions
		• The NIH Reentry into Biomedical Research Careers program supplements existing NIH research grants to support full- or part-time research by women or men returning to the scientific workforce. The program is designed to bring a scientist's existing research skills and knowledge up-to-date so that by the end of the supplement period, the scientist will be prepared to apply for a career development award, a research award, or another form of independent research support. • NSF reported that under the Career Life Balance (CLB) Initiative, NSF awarded several grants to institutions of higher education to conduct and disseminate findings of research on effectiveness of institutional strategies to promote career-life balance. For example, award # 1446406 "On Ramps" to Full Professor: Institutional Support for Post-Family Leave Faculty Research Reintegration aims to develop and disseminate a post-family leave reintegration model to accelerate post-family leave female faculty progression in STEM.
Provide support for project personnel while researchers are on family leave	Establish policies that allow grants to cover salaries of project personnel to allow research to continue while researchers are on family leave.	• USDA-NIFA reported that its award terms and conditions allow grantee institutions to appoint a temporary Project Director if the Project Director or other researcher is on leave. • DOD AT&L reported that there are government-wide requirements that allow research to continue while the principal investigator/project director is on family leave, as long as the plans for continuing the research during the absence of that are acceptable to the awarding agency. For all other staff, as long as the research progresses, DOD does not specify how the work should be accomplished.

Action	Description	Summaries of Selected Agency Responses to Identified List of Actions
		• DOE reported that an institution may request a supplement to a research grant to cover the extra help. • NASA noted that its Research Opportunities in Space and Earth Sciences (ROSES) solicitation accommodates all reasonable requests to re-phase ROSES awards (e.g., for family or medical leave), and that in early career and ROSES programs, NASA is willing to accommodate those who have been delayed in their careers due to family leave. NASA officials also noted there is currently nothing in regulation or policy that prevents grantees from covering the salaries of temporary staff. If a Principal Investigator is absent for a period of time, the university has the responsibility to ensure the project continues and can use the grant funding for this purpose. NASA awards include terms and conditions regarding the absence of a Principal Investigator from the project. • NIH grant awards (with the exception of National Research Service Awards) allow for reimbursement of actual, allowable costs incurred for parental leave and additional technical support provided such costs are incurred under formally-established institutional policies that are consistently applied regardless of the source of support. • NSF reported that it issued frequently asked questions related to career-life balance. These FAQs specifically address issues regarding use of grant funds to cover technicians to allow research to continue while researchers are on family leave.

Source: GAO analysis of literature, expert input, and federal agency interviews and data. | GAO-16-14.

Table 16. Possible Actions We Identified through a Literature Review and Expert Consultation Related to Overseeing the Research Proposal Review Process for Federal STEM Grant-making Agencies

Action	Description	Summaries of Selected Agency Responses to Identified List of Actions
Conduct blind reviews of grant applications	Institute full or partial blind reviews when reviewing proposals to mask applicants' identities in the early stages of review, before an applicant's qualifications need to be assessed.	• Department of Agriculture-National Institute for Food and Agriculture (USDA-NIFA) officials reported that NIFA does not conduct blind reviews. NIFA officials stated that instituting a blind review process would be difficult because the Project Director's qualifications are a key consideration during the proposal review process. • Department of Defense (DOD) officials did not think blind reviews would be possible given DOD's evaluation criteria, which includes an assessment of researcher capability. • Department of Energy's (DOE) Office of Nuclear Energy (NE) instituted a partial blind review beginning in 2010 wherein applicants' identities are not revealed until after the panel has assessed and scored the applicants' initial technical proposals. NE officials reported this process has been successful in diversifying its pool of grantees, such as to include younger researchers, female researchers, and researchers new to NE. • DOE's Office of Science officials reported that the Office of Science's regulations for proposal review would not support conducting blind reviews since one of its required merit review criteria asks about the competency of the applicant's personnel. • National Aeronautics and Space Administration (NASA) officials reported that it does not currently conduct blind reviews, and to do so would be impossible given the evaluation criteria to evaluate the qualifications and capabilities of the research team under its merit review process.

Action	Description	Summaries of Selected Agency Responses to Identified List of Actions
		• National Institutes of Health (NIH) does not currently conduct blind reviews though NIH reported that it launched two America COMPETES Act challenges to help identify new methods to detect bias in peer review and strategies to strengthen fairness and impartiality in peer review. • National Science Foundation (NSF) reported that the NSF merit review process is not currently structured for blind reviews. Per the merit review criteria as established by the National Science Board, among the factors that proposal reviewers should consider is "how well qualified is the individual, team or organization to conduct the proposed activities."
Seek diversity among proposal reviewers	Establish goals and objectives for attaining gender diversity among proposal reviewers.	• USDA-NIFA reported that it tracks the race, ethnicity and gender of individuals who sit on its panels with a goal of increasing racial, ethnic, and gender diversity of panelists. • While DOD does not currently set goals for attaining gender diversity among proposal reviewers, DOD Acquisition, Technology and Logistics (AT&L) officials suggested DOD could establish a baseline measurement of its current proposal reviewers and use this to develop strategies to involve more women as proposal reviewers. • DOE Office of Science reported that its senior management encourages program managers to take all types of diversity into account when choosing proposal reviewers. In addition, Committees of Visitors review its internal processes every three years, and look at diversity in the reviewer pool. • NASA reported that it does not currently set goals for attaining gender diversity among proposal reviewers, noting this is a policy decision that would need to be made by Agency senior leadership, acting in concert with lead technical offices, such as the Office of the Chief Scientist, and the grant awarding Mission Directorates.

Table 16. (Continued)

Action	Description	Summaries of Selected Agency Responses to Identified List of Actions
		• NIH reported it welcomes individuals with diverse backgrounds to consider joining its review groups so that the panels are diverse with respect to geographic representation, gender, race, and ethnicity. • NSF administrative guidelines include the instruction that, when selecting reviewers, special attention should be paid to obtaining qualified persons from underrepresented groups, such as ethnic minorities, women, and individuals with disabilities. Policies on the selection of reviewers are described in the training provided to all incoming program officers. Division directors are given the responsibility of monitoring the diversity and expertise of reviewers selected by the programs in their division. Annually, NSF publishes aggregate data on the diversity of reviewers used in the prior year. One way NSF is trying to increase diversity among reviewers is to make greater use of technology to enable reviewers to participate remotely in review panels in order to reduce barriers to participation. The virtual panelist project is a three-year pilot effort and there is a NSF internal effort, led by the Chief Technology Officer, to review the impact of the use of virtual review panels.
Conduct unconscious bias training for staff[a]	Conduct workshops about possible gender bias in federal grant-making processes and consider coordinating with other federal agencies to leverage common resources and lessons learned.	• USDA-NIFA did not comment on this action. • DOD reported that some DOD components have held brownbag lunchtime sessions on unconscious bias. AT&L noted DOD could also leverage experiences of agencies in the Broadening Participation Interagency WorkingGroup, NIH's Advisory Committee report on Diversity in the Biomedical Workforce, and the Subcommittee on Peer Review.

Action	Description	Summaries of Selected Agency Responses to Identified List of Actions
		• DOE reported having co-sponsored several gender equity workshops with NIH and NSF, though none focused on gender bias in federal grant-making. • NASA reported that an internal agency training module on addressing unconscious bias in employment processes is under development. • NIH has trained its top leaders in bias-awareness. • NSF developed an implicit bias training program at the NSF Academy for NSF internal program officers to review/take and train them on the implicit bias. NSF has also created slides that grant proposal reviewers can reference that will help introduce them to the issues related to implicit bias. In addition, NSF participates in exchanges of information/best practices regarding federal agency research administration in both formal and informal settings, e.g., the National Science and Technology Council.

Source: GAO analysis of literature, expert input, and federal agency interviews and data. | GAO-16-14

[a]This action would apply to both agency staff and external review panelists who review and make decisions about funding for research proposals.

Table 17. Possible Actions We Identified Through a Literature Review and Expert Consultation Related to Funding and Assisting Academic Institutions for Federal STEM Grant-making Agencies

Action	Description	Summaries of Selected Agency Responses to Identified List of Actions
Fund additional research to address the representation of women in STEM	Expand support for research on the obstacles to diversity in STEM research fields, as well as evaluations of interventions that address these obstacles.	• Department of Agriculture-National Institute of Food and Agriculture (USDA-NIFA) reported that it does not receive appropriated funding to support research on the obstacles to diversity in STEM research fields.

Table 17. (Continued)

Action	Description	Summaries of Selected Agency Responses to Identified List of Actions
		However, NIFA reported that it has five representatives on the Federal Commission on STEM Working Groups, including the Broadening Participation Working Group. These interagency working groups are discussing obstacles to diversity in STEM fields and strategies to address those obstacles. • Department of Defense (DOD) does not directly fund such research, however DOD's Acquisition, Technology and Logistics (AT&L) officials reported that DOD funds the STEM Diversity Campaign, which conducts research on obstacles in STEM diversity and DOD could leverage experiences of agencies in the Broadening Participation Interagency Working Group. • Department of Energy (DOE) Office of Science reported that it has one active grant to support research on the recruitment, retention, and promotion of women in the chemical sciences. • National Aeronautics and Space Administration (NASA) does not currently fund such research, though officials reported that the feasibility of this is something that Agency senior leadership, the Mission Directorates, the Office of Procurement, as well as the NASA Office of Education, would have to consider, with advice from the Office of the General Counsel. This could only be accomplished if it is part of the authorizing legislation and fits into NASA's mission. • National Institutes of Health (NIH) funded 14 grants in October 2009 totaling $16.8 million over 4 years to further research on the obstacles facing women and interventions that promote and support the careers of women in biomedical and behavioral science and engineering.

Action	Description	Summaries of Selected Agency Responses to Identified List of Actions
		NIH also supports the National Research Mentoring Network, a nationwide consortium to enhance the training and career development of individuals from diverse backgrounds who are pursuing biomedical research careers through the development of best practices for mentoring, by providing training opportunities for mentors, and by providing networking and professional opportunities for mentees. • National Science Foundation (NSF) provides research funding to universities through its ADVANCE program to address various aspects of STEM academic culture and institutional structure that may differentially affect women faculty and academic administrators.
Host workshops to present research findings	Host workshops with grantees and institutional stakeholders to present and discuss research findings and develop potential interventions.	• USDA-NIFA did not report on NIFA-supported workshops at the graduate level. • DOD reported it has not taken this action, but could collaborate with agencies such as NIH and NSF to see how they have successfully implemented, or plan to implement, such workshops. • DOE reported that it has co-sponsored several such workshops with the NIH and NSF, for example, on gender equity in physics, materials science, and chemistry. • NASA officials reported that NASA is planning a series of virtual sessions with grantees, which will address topics such as family friendly policies in STEM and recruitment and retention of female students and faculty. • NIH officials reported that its Working Group on Women in Biomedical Careers has held a workshop for its grantees to discuss data, findings, and next steps related to their research on the causal factors and interventions that promote and support the careers of women in biomedical and behavioral science and engineering.

Table 17. (Continued)

Action	Description	Summaries of Selected Agency Responses to Identified List of Actions
		• NSF did not report on NSF-supported workshops at the graduate level.
Encourage grant-recipient institutions to conduct unconscious bias workshops	Encourage grant-recipient universities to conduct evidence-based workshops on unconscious bias for faculty department chairs, professors, deans, and administrators at all levels of the STEM pipeline.	• USDA-NIFA reported this is not feasible due to current terms and conditions within research grants. • DOD reported its relationship with universities does not extend to encouraging them to conduct such workshops. • DOE reported co-sponsoring several such workshops with NIH and NSF. • NASA reported having recommended this to grantee institutions in its Title IX compliance reports. NASA has also developed an unconscious bias training tool, available online, for universities to use, after a literature review showed a need for unconscious bias training. • NIH reported that its National Institute of General Medical Sciences funds a research grant project which aims to: 1) develop and implement training for research mentors, faculty, and future faculty about implicit bias; 2) implement "Train the Trainer" workshops for previous participants; and 3) disseminate breaking the cycle of bias curriculum and training materials. • NSF officials reported that the ADVANCE program supports evidence-based workshops on bias awareness. In addition to helping ADVANCE-funded institutions to recognize and address gender bias in STEM via institutional-tailored workshops, the ADVANCE program is supporting the Association of Women in Science (AWIS) to leverage a national platform to promote change through the reduction of implicit gender bias. Additionally, the Science of Broadening Participation portfolio contributes to the scientific literature on implicit bias in STEM.

Source: GAO analysis of literature, expert input, and federal agency interviews and data. | GAO-16-14

End Notes

[1] Pub. L. No. 92-318, tit. IX, § 901, 86 Stat. 235, 373, codified at 20 U.S.C. §§ 1681-1688.

[2] National Science Foundation, National Center for Science and Engineering Statistics. *Women, Minorities, and Persons with Disabilities in Science and Engineering: 2015* (Arlington, VA: Special Report NSF 15-311). Available at http://www.nsf.gov/statistics/wmpd/.

[3] This figure is reported annually by the National Science Foundation's National Center for Science and Engineering Statistics and includes the total amount of federal obligations for research at universities and colleges in the fields of computer science and mathematics, engineering, environmental sciences, life sciences, and physical sciences. The amount may include contract funding in addition to grants and cooperative agreements. However, the focus of this report is on grant funding awarded to universities for STEM research.

[4] Title IX applies to all entities that receive federal funding for education programs or activities. This includes but is not limited to universities that receive federal STEM research grant funding.

[5] GAO, *Women in STEM Research: Federal Agencies Differ in the Data They Collect on Grant Applicants*, GAO-15-291R (Washington, D.C.: Mar. 17, 2015).

[6] In FY 2014 NIFA, a component of USDA, was responsible for nearly 90 percent of USDA's STEM research funding provided to universities. Therefore, we focused on NIFA in this report.

[7] In FY 2014 NIH, a component of the Department of Health and Human Services (HHS), was responsible for administering over 98 percent of the department's research budget. Note we focused only on NIH's extramural research funding in this report.

[8] Other definitions of STEM may include healthcare fields such as health practitioners and technicians, STEM educators, and researchers in other fields such as the social sciences, psychology, and multidisciplinary fields. In general, data show women are better represented in healthcare and other STEM fields.

[9] National Science Foundation, National Center for Science and Engineering Statistics. *Women, Minorities, and Persons with Disabilities in Science and Engineering: 2015.*

[10] In some agencies and components, fewer than five years of reliable data were available; in these instances, we analyzed available data from the most recent fiscal years during this time period.

[11] The SDR asks respondents whether their work is supported by federal funding and if so, from which federal agency, including DOD, DOE, NIH, and NSF. Therefore, we were unable to analyze USDA-NIFA and NASA using this dataset.

[12] GAO, *Standards for Internal Control in the Federal Government*, GAO/AIMD-00-21.3.1 (Washington, D.C.: November 1999). GAO recently revised and reissued Standards for Internal Control in the Federal Government, with the new revision effective beginning with fiscal year 2016. GAO-14-704G (Washington, D.C.: September 2014).

[13] GAO/AIMD-00-21.3.1.

[14] We selected experts who were knowledgeable about the federal role in increasing STEM diversity, including contributors to the academic literature, academic department chairs, retired federal officials, and leaders in industry and professional organizations. See appendix I for further information on our methodology for selecting experts.

[15] For the purposes of this report, we will use the term "universities" to refer to all institutions of higher education, including colleges. In addition, the entities we reviewed within each agency vary and may be referred to as offices, agencies, commands, or components. For the

purposes of this report, we will refer to these separate entities within a larger agency as "components."

[16] Officials at one agency component told us that contact between a potential applicant and a program officer was prohibited once a funding announcement had been officially posted.

[17] In 2002, Title IX of the Education Amendments of 1972 was renamed the Patsy Takemoto Mink Equal Opportunity in Education Act. Pub. L. No. 107-255, 116 Stat. 1734 (2002). For purposes of this report, we refer to this Act as Title IX.

[18] Specifically, with certain exceptions, Title IX prohibits sex discrimination in education programs or activities receiving any form of federal financial assistance. 20 U.S.C. § 1681(a).

[19] 20 U.S.C. § 1682. Education and HHS's Title IX regulations derive from the Department of Health, Education, and Welfare (HEW) Title IX regulations published in 1975 (40 Fed. Reg. 24,137, June 4, 1975). In 1979 the Department of Education was created and HEW was renamed the Department of Health and Human Services. Education's Title IX regulations are now codified at 34 C.F.R. Part 106 and HHS's regulations are codified at 45 C.F.R. Part 86. USDA published regulations in 1979 (44 Fed. Reg. 21,610, April 11, 1979, codified at 7 C.F.R. Part 15a) and DOE published regulations in 1980 (45 Fed. Reg. 40,514, June 13, 1980, codified at 10 C.F.R. Part 1042). A Title IX Final Common Rule was published in 2000 to implement regulations for 20 agencies, including DOD, NASA, and NSF (65 Fed. Reg. 52,858, Aug. 30, 2000). In 2001, DOE replaced its regulations with the provisions of the Common Rule (66 Fed. Reg. 4630, Jan. 18, 2001). Title IX regulations for all of the agencies in our review are substantially the same. For purposes of this report any reference to a regulatory requirement is applicable to all six agencies we reviewed.

[20] See, for example, HHS regulations at 45 C.F.R. §§ 80.7(a), 86.71. The specific requirement to conduct periodic compliance reviews is generally found in agencies' regulations enacted pursuant to Title VI of the Civil Rights Act of 1964, as amended (42 U.S.C. § 2000d, et seq.). However, as stated in the preamble to the common rule, "[f]or those agencies that have regulations to enforce Title VI, such procedures are adopted and referenced. Titles VI and IX address discrimination in federally assisted programs and have identical statutory enforcement schemes. The administrative enforcement procedures in Title VI regulations [including the requirement to conduct periodic compliance reviews of funding recipients] are virtually identical among the participating agencies, and differences are minor." 65 Fed. Reg. 52,858, 52,860 (Aug. 30, 2000). While all six federal agencies we reviewed are required to conduct periodic compliance reviews of their recipients, the NASA Reauthorization Act of 2005 further required NASA to conduct Title IX compliance reviews of at least two grantees per year. Pub. L. No. 109- 155, § 619, 119 Stat. 2895, 2935 (codified at 51 U.S.C. § 40909). Similarly, the America COMPETES Act contained a Sense of Congress provision that DOE should also conduct Title IX compliance reviews of a minimum of two grantees per year. Pub. L. No. 110-69, § 5010, 121 Stat. 572, 620 (2007).

[21] See, for example, NSF regulations at 45 C.F.R. §§ 611.7(b), 618.605. In addition, an individual alleging discrimination on the basis of sex generally may also choose to file a discrimination complaint under his or her university's grievance system.

[22] While Title IX applies to all recipients of federal funding for educational programs or activities including, but not limited to, grantees, for the purposes of this report, which pertains to universities receiving research grant funds from federal agencies, we will refer to recipients as "grantees."

[23] Exec. Order No. 12250 (1980). Specifically, DOJ's Civil Rights Division is responsible for enforcing federal statutes prohibiting discrimination of protected classes, including Title IX.

[24] NSF, National Center for Science and Engineering Statistics, *Women, Minorities, and Persons with Disabilities in Science and Engineering: 2015*.

[25] 2013 is the most recent year for which these data are available. See National Science Foundation, National Center for Science and Engineering Statistics. *Women, Minorities, and Persons with Disabilities in Science and Engineering: 2015*.

[26] Ibid.

[27] See Ceci, Stephen J., and Wendy M. Williams. "Understanding Current Causes of Women's Underrepresentation in Science." *Proceedings of the National Academy of Sciences* 108, no. 8 (2011): 3157-3162; see also: Goulden, Marc. Mary Ann Mason, and Karie Frasch. "Keeping Women in the Science Pipeline." *The ANNALS of the American Academy of Political and Social Science* November 2011 *vol. 638 no. 1* 141-162. Goulden, et al., provide evidence that women drop out of the academic track in disproportionate numbers due to family formation. In addition, studies have shown that female physicians working in non-academic careers work more hours than scientists but do not experience their most intense period of professional competition during child bearing years, allowing for somewhat more flexibility (see Adamo, Shelley A. "Attrition of Women in the Biological Sciences: Workload, Motherhood, and Other Explanations Revisited." *BioScience (2013) 63 (1): 43-48)*. Another study finds female academics report one barrier to taking advantage of maternity leave policies is an inability to stop work, particularly on grant-funded projects (see Villablanca, Amparo C. et al. "Career Flexibility and Family-Friendly Policies: An NIH-Funded Study to Enhance Women's Careers in Biomedical Sciences." *Journal of Women's Health*. 2011 Oct; 20(10): 1485–1496).

[28] Ceci and Williams (2011).

[29] National Academy of Sciences, National Academy of Engineering, and Institute of Medicine, Committee on Maximizing the Potential of Women in Academic Science and Engineering, *Beyond Bias and Barriers: Fulfilling the Potential of Women in Academic Science and Engineering* (Washington, D.C.: 2006).

[30] Goulden, Mason, Frasch (2011).

[31] Mason, Mary Ann, and Marc Goulden. "Marriage and Baby Blues: Redefining Gender Equity in the Academy." *The Annals of the American Academy of Political and Social Science* 596, no. 1 (2004): 86-103.

[32] *Gender Differences at Critical Transitions in the Careers of Science, Engineering, and Mathematics Faculty*. (Washington, DC: National Academies Press, 2010).

[33] For example, one finding was that women who applied in one year were less likely to apply in subsequent years. See Susan D. Hosek, et al, *Gender Differences in Major Federal External Grant Programs* (Santa Monica: RAND Corporation, 2005). See also Pohlhaus, Jennifer Reineke, et al. "Sex Differences in Application, Success, and Funding Rates for NIH Extramural Programs." *Academic Medicine* 86.6 (2011): 759 and Ley, Timothy J. and Barton H. Hamilton, "The Gender Gap in NIH Grant Applications", *Science*, vol. 332 (2008).

[34] In the report *Beyond Bias and Barriers: Fulfilling the Potential of Women in Academic Science and Engineering* (Washington, D.C.: The National Academies Press, 2006), the authors note that although overt discrimination toward women is no longer considered socially acceptable "people may still hold prejudiced attitudes, stemming in part from the US history of overt sex and racial prejudice..." Additionally, research has shown that "even individuals who espouse a belief of gender equity and equality may harbor implicit biases about gender and, hence, negative gender stereotypes about women and girls in science and math". See Hill, Catherine, Christianne Corbett, and Andresse St Rose. *Why So Few? Women in*

Science, Technology, Engineering, and Mathematics. American Association of University Women. Washington, DC. See also Moss-Racusin, Corinne A. et al. "Science faculty's subtle gender biases favor male students." *Proceedings of the National Academy of Sciences* 109, no. 41 (2012): 16474-16479, and Hyde, Janet S., and Janet E. Mertz. "Gender, Culture, and Mathematics Performance" *Proceedings of the National Academy of Sciences* 106, no. 22 (2009): 8801-8807.

[35] See *Beyond Bias and Barriers* (2006).

[36] *Beyond Bias and Barriers* (2006). See also Nolan, Susan A., et al. "Training and Mentoring of Chemists: A study of Gender Disparity." *Sex Roles* 58, no. 3-4 (2008): 235- 250., McDowell, John M., Larry D. Singell, Jr. and Mark Stater. "Two to Tango? Gender Differences in the Decisions to Publish and Coauthor." *Economic inquiry* 44, no. 1 (2006): 153-168., and Blau, Francine D., et al. *Can Mentoring Help Female Assistant Professors? Interim Results From a Randomized Trial*. No. w15707. National Bureau of Economic Research, 2010.

[37] Hill, et al. *Why So Few? Women in Science, Technology, Engineering, and Mathematics*. (2010) and *Beyond Bias and Barriers* (2006). See also GAO, *Gender Issues: Women's Participation in the Sciences Has Increased, but Agencies Need to Do More to Ensure Compliance with Title IX*. GAO-04-639 (Washington, D.C.: Jul. 22, 2004).

[38] The National Science Foundation Authorization Act of 2002 (Pub. L. No. 107-368, § 18(c), 116 Stat. 3034, 3061), directed NSF to sponsor a study to "examine differences in amounts requested and awarded, by gender, in major Federal external grant programs". See Susan D. Hosek, et al, *Gender Differences in Major Federal External Grant Programs* (Santa Monica: RAND Corporation, 2005).

[39] See GAO-15-291R. Studies focusing on NIH and NSF, including the RAND study, have not faced these types of data limitations. Furthermore, both NIH and NSF regularly report on their grant-making programs and outcomes. See, for example, National Institutes of Health, *NIH Data Book*, http://report.nih.gov/nihdatabook/ and National Science Foundation, *Report to the National Science Board on the National Science Foundation's Merit Review Process Fiscal Year 2014* (NSB-2015-14: May 2015).

[40] We defined the success rate as the ratio of awards made to proposals reviewed (specifically, those with a final disposition), excluding revisions in the same fiscal year. For more on how the success rates in this report were calculated, please see appendix I. Success rates in this report are aggregate over time and award type and may differ from success rates reported elsewhere, such as those broken down by award type.

[41] At NIH, NSF, and USDA-NIFA, researcher sex and other demographic characteristics are collected separately from specific proposal information and are not available to proposal reviewers or used in the grant decision-making process. Rather, these data are used for internal analysis of how different demographic groups fare in the grant process.

[42] See National Institutes of Health, NIH Data Book, http://report.nih.gov/nihdatabook/, National Science Foundation, Report to the National Science Board on the National Science Foundation's Merit Review Process Fiscal Year 2014, NSB-2015-14 (May 2015), and USDA-NIFA Agriculture and Food Research Initiative, AFRI 2013 Synopsis, http://nifa.usda.gov/afri-2013-synopsis.

[43] NIH Office of Extramural Research, 2013-2014 OER Report, http://grants.nih.gov/grants/2013_OER_Report.pdf.

[44] This *FY 2014 Report on the NSF Merit Review Process* responds to a National Science Board (NSB) policy, endorsed in 1977 and amended in 1984, requesting that the NSF Director submit an annual report on the NSF merit review process.

[45] See, for example, USDA-NIFA's data gateway online: http://nifa.usda.gov/data.

[46] To calculate success rates for DOD and DOE (where sex is not self-reported by applicants) we used a methodology to determine the sex of applicants and awardees by matching first names to the Social Security Administration's (SSA) baby name database, a strategy also used by RAND Corporation in its 2005 report (see Susan D. Hosek, et al, *Gender Differences in Major Federal External Grant Programs* (Santa Monica: RAND Corporation, 2005). To test the accuracy, sensitivity, and robustness of this technique, we tested our name-matching methodology against an agency dataset with self-reported gender and also conducted additional robustness checks. See appendix I for further details.

[47] The Office of Naval Research is responsible for approximately 29 percent of DOD's funding for STEM research at universities. The Army's research components together fund about 17 percent of DOD's total. The Defense Threat Reduction Agency and other small components of DOD combined make up approximately 7 percent of DOD's total. There were insufficient data to analyze success rates at the Defense Advanced Research Projects Agency, DOD's largest research grant funding component.

[48] For example, at the Advanced Research Projects Agency-Energy in fiscal year 2010, only 4 of 28 awards were made to women. Given the relatively few female applicants in this year, the success rate was 25 percent for women and 12.5 percent for men. Similarly, in fiscal year 2011, the Office of Energy Efficiency and Renewable Energy awarded 4 of 27 awards to women, with success rates of 6.9 and 5.7 percent for women and men, respectively. Because of this, analyzing success rates alone obscures the fact that women still receive fewer grant awards relative to men. The small numbers at these components also make our analysis particularly sensitive to the assumptions in our gender name match methodology. In contrast, DOE reported that in fiscal year 2012, the Office of Science awarded 80 percent of DOE's total grant funding to universities for STEM research and funded over 1,000 of the new, renewal, and supplemental proposals it received. In the years we reviewed, the Office of Science funded between approximately 740 and 1,200 new, renewal, and supplemental proposals each year, with an average of around 980 funded proposals per year. See appendix I for further details.

[49] Our methodology for calculating success rates required linking proposals to any resulting award to ensure that any awards counted in the numerator of the success rate had a corresponding proposal in the denominator of the success rate (along with the unsuccessful proposals). Using this method to calculate success rates, we could not calculate a success rate for NASA.

[50] It is impossible to say with any degree of certainty whether the 1 or 2 percentage point differences in success rates at DOD's Army Research Office and Defense Threat Reduction Agency are small enough to ignore, or whether the 6 and 7 percentage point differences at DOD's Office of Naval Research and DOE's Office of Science are too large to be ignored. The determination of a threshold for action is a policy decision for which our analysis provides limited information.

[51] As noted in GAO-15-291R, while NIH, NSF, and USDA-NIFA collect data on demographic and educational characteristics of applicants, NASA and many components of DOD and DOE do not. Therefore, we could not rely on agency data to explore the effect of researcher characteristics on the likelihood of funding success.

[52] A fourth component at DOD, the Air Force Office of Scientific Research, began tracking proposals, awards, and grants in one system in 2014. We determined data from this component prior to 2014 were not sufficiently reliable for inclusion in this study.

[53] GAO-15-291R.

[54] While officials at NIH and NSF, the two largest federal grant-making agencies, told us that potential applicants are encouraged to initiate a discussion of their research proposal ideas with their program officer prior to submitting full proposals at these two agencies, it is generally not a requirement.

[55] GAO-15-291R.

[56] DOE further noted that any legal concerns it has with the collection of certain PII data on applicants does not extend to the collection of grant life cycle data, which it noted is useful.

[57] GAO/AIMD-00-21.3.1.

[58] Officials from each of the six grant-making agencies we reviewed told us that their agency has received few, if any, Title IX complaints from students, faculty, and staff at universities it funds for STEM research, perhaps because individuals may not be aware that grant-making agencies have authority to receive and investigate Title IX complaints or that individuals would be more likely to use the university's own process for receiving and investigating complaints. Some of these agencies, including NASA and DOE, periodically provide brochures to grantee universities outlining their agency's authority to receive and investigate Title IX complaints, but officials at these agencies told us that this outreach has not resulted in any Title IX complaints against their university grantees being filed with their agency.

[59] During one desk review in 2010, NASA found one university grantee to be in noncompliance because it did not have a designated Title IX coordinator. However, before NASA issued a letter of non-compliance, it informed the school and the school reported it had named a qualified Title IX coordinator. As a result, NASA did not issue a formal noncompliance letter.

[60] In 2004, we reviewed Title IX compliance efforts at DOE, NASA, and NSF and found that these agencies had not conducted Title IX compliance reviews. We made recommendations to each agency to begin conducting these reviews, see GAO-04-639. In response to our recommendation and consistent with subsequent legislation, NASA has conducted two onsite compliance reviews each year since 2006 in addition to 15 total Title IX compliance desk reviews. According to officials, DOE and NSF also completed one joint compliance review in 2005. While DOE said it has conducted 15 compliance reviews since December 2005, NSF officials told us that its Title IX compliance review program was on hold from 2006 to 2014. According to officials, in fall 2014, NSF restarted its compliance review program and, in May 2015, NSF conducted one joint compliance review with DOE. According to NSF officials, these two agencies planned to conduct one more joint review in September 2015.

[61] In 2009, NASA, for instance, compiled and published a list of promising practices it identified while conducting Title IX compliance reviews. See NASA, *Title IX & STEM: Promising Practices for Science, Technology, Engineering, & Mathematics* (Washington, D.C.: June 2009). NASA has also published guidance to assist universities wishing to conduct self-evaluations of their efforts to comply with Title IX and enhance gender equity at their institutions. See NASA, *Title IX & STEM: A Guide for Conducting Title IX Self-Evaluations in Science, Technology, Engineering, and Mathematics Programs* (Washington, D.C.: June 2012).

[62] The HHS departmental Office for Civil Rights is responsible for ensuring Title IX compliance at all of HHS's funding recipients, including investigating complaints and requiring assurances of compliance at NIH's grantees. Therefore, in this section of our report, we will refer to HHS rather than NIH when discussing Title IX compliance activities.

[63] According to officials from DOD's Office of Diversity Management and Equal Opportunity (ODMEO), each DOD grant-making component is responsible for conducting Title IX compliance reviews of their own grantees.

[64] Department of Defense, *Directive Number 5500.11: Nondiscrimination in Federally Assisted Programs* (Washington, D.C.: May 27, 1971, certified current as of Nov. 21, 2003).

[65] The one exception at DOD was the Defense Threat Reduction Agency, where officials responsible for civil rights compliance told us they made one Title IX compliance review visit to a grantee in July 2015.

[66] Officials noted that the agency has overlapping Title IX compliance responsibility with the Department of Education over university grantees, but that HHS has primary or sole responsibility for enforcing compliance of certain nondiscrimination provisions in the Affordable Care Act, for example.

[67] While not specifically commenting on the compliance efforts of any one agency, DOJ has generally noted that federal agencies should not assume that the absence of complaints from a particular segment of beneficiaries or those in a particular area means that problems do not exist. HHS OCR officials stated they were only able to verify whether there were any Title IX complaints against NIH grantees since the year 2000.

[68] HHS officials also told us it has also developed a civil rights training in conjunction with the American Association of Medical Colleges and has provided this training to medical school faculty and students at numerous conferences since 2007. However, this training is much broader than Title IX and only mentions sex discrimination in passing.

[69] As we noted earlier in our report, this figure includes the total NIH funding for research provided to universities in FY 2014 in the STEM fields of computer science and mathematics, engineering, environmental sciences, life sciences, and physical sciences, consistent with our focus on core STEM fields. The scope of this review does not include funding provided to universities in other disciplines sometimes included in the definition of STEM, including psychology and the social sciences, or STEM education programs.

[70] Department of Justice, *Investigation Procedures Manual for the Investigation and Resolution of Complaints Alleging Violations of Title VI and Other Nondiscrimination Statutes* (Washington, D.C.: September 1998).

[71] Fed. Reg. Vol. 65, No. 169, *Nondiscrimination on the Basis of Sex in Education Programs or Activities Receiving Federal Financial Assistance; Final Common Rule* (Washington, D.C: Aug. 30, 2000) and Department of Justice, *Title IX Legal Manual* (Washington, D.C.: Jan. 11, 2001).

[72] Ex. Order No. 12250, § 1-206.

[73] DOJ noted that STEM grant-making agencies have independent obligations to conduct compliance reviews and that these six STEM grant-making agencies could also initiate such Title IX information sharing on their own.

[74] According to DOJ officials, DOJ is a member of a STEM working group created through the White House Council on Women and Girls, which began meeting in January 2014 and includes federal agencies that work on STEM issues. DOJ stated that at a meeting of the working group, it reminded agency officials in attendance that DOJ is available to assist them with Title IX compliance issues. However, Executive Order 13506, which establishes the White House Council, provides a broadly focused mission for this group and does not specifically mention Title IX.

[75] DOJ reported that it currently focuses its resources on the following Title IX activities: conducting compliance reviews and investigations, ensuring compliance with consent decrees and settlements, reviewing agencies' Title IX guidance, providing technical

assistance to other federal agencies upon request—including in their Title IX compliance reviews—and participating in the White House Task Force to Protect Students from Sexual Assault.

[76] This section of the report focuses on possible actions agencies could take beyond the additional data collection and Title IX enforcement activities discussed in the prior sections of this report.

[77] Institute for Women's Policy Research, *Accelerating Change for Women Faculty of Color in STEM: Policy, Action, and Collaboration.* (Washington, D.C.: Institute for Women's Policy Research, 2013). We did not evaluate the efficacy of these actions and this report does not establish whether they will achieve their intended goals or whether they would be suited to each agency's particular organizational context.

[78] For detailed descriptions of these actions and a summary of agency responses about each action, see appendix III.

[79] Hill, et al. *Why So Few? Women in Science, Technology, Engineering, and Mathematics.* (2010) and *Beyond Bias and Barriers* (2006). M. Goulden, M.A. Mason, and K. Frasch. "Keeping Women in the Science Pipeline." *Annals, AAPSS*, vol. 638 (2011).

[80] GAO, *Diversity Management: Expert-Identified Leading Practices and Agency Examples,* GAO-05-90 (Washington, D.C.: Jan. 14, 2005).

[81] University of Michigan, *Year-End Report for ADVANCE Project* (University of Michigan: 2002).

[82] Xie, Yu and Kimberlee A. Shauman. *Women in Science: Career Processes and Outcomes.* (Cambridge, MA: Harvard University Press, 2003). See also Goulden, Mason, Frasch (2011).

[83] Goulden, Mason, Frasch (2011).

[84] *Beyond Bias and Barriers* (2006).

[85] NIH officials reported that NIH grant awards (with the exception of National Research Service Awards) allow for reimbursement of actual, allowable costs incurred for child care provided such costs are incurred under formally-established institutional policies that are consistently applied regardless of the source of support. In 2013, OMB streamlined the federal government's guidance on grants management, stating that it included provisions in its final rule to encourage non-federal entities to have family-friendly policies. Uniform Administrative Requirements, Cost Principles, and Audit Requirements for Federal Awards, 78 Fed. Reg. 78,590, 78,593 (Dec. 26, 2013). For example, included in this rule was a change that allows grantees to be reimbursed for costs of identifying, but not providing, locally available dependent care options while attending conferences necessary and reasonable for successful performance under the award. 78 Fed. Reg. 78,590, 78,650. In addition, NIH and NSF currently lead an interagency effort to develop more specific guidance, including a Frequently Asked Questions document on family-friendly policies, for federal awards and at non-federal entities (e.g., university grantees) through the Research Business Models Working Group of the Committee on Science of the National Science and Technology Council.

[86] Moss-Racusin, et al. "Science faculty's subtle gender biases favor male students." (2012).

[87] Åslund, Olof and Oskar Nordström Skans. "Do Anonymous Job Application Procedures Level the Playing Field?" *Industrial and Labor Relations Review,* vol. 65, no. 1, (January 2012) and Board of Editors. "Preferential Treatment: Good Intentions are Not Enough to End Racial and Gender Bias." *Scientific American.* (October 2014).

[88] Phillips, Katherine W. "How Diversity Works." *Scientific American.* (October 2014).

[89] Research has found that women in biomedical research have applied for NIH grants in disproportionately low numbers compared with their male counterparts, in part due to high levels of attrition among women during mid-career. See Ley and Hamilton, (2008), Ginther, et al, (2011), and Pohlhaus and Reineke, et al (2011).

End Notes for Appendix I

[1] According to 2014 NSF data, together, these agencies funded 90 percent of the federal government's investment in basic and applied research in STEM fields from fiscal year 2012 through 2014.

[2] In FY 2014, NIFA, a component of USDA, was responsible for 90 percent of USDA's STEM research funding provided to universities. Therefore, we focused on NIFA in this report.

[3] In FY 2014, NIH, a component of the Department of Health and Human Services (HHS), was responsible for administering nearly 99 percent of the department's research budget. Therefore, we focused on NIH in this report.

[4] In this report, we included the core STEM fields of engineering, life sciences, physical sciences, computer and information technology, and mathematics and statistics.

[5] GAO/AIMD-00-21.3.1.

[6] The SDR is a biannual survey with a sample size of approximately 47,000 from a population of approximately 840,000 in 2013. The population of the SDR consists of all non-institutionalized, non-terminally ill individuals who were younger than 76 years of age, who received a science, engineering, or health research doctorate from a U.S. academic institution. It asks respondents whether their work is supported by federal funding and, if so, from which federal agency, including DOD, DOE, NIH, and NSF. The SDR collects information on individuals' education, careers, demographic characteristics, and family and marital situations, allowing us to create a statistical model that predicts the odds of federal funding for women versus men; 2013 was the most recent year of SDR data available for our analysis.

[7] For additional information on the data elements collected by each of these agencies and components, see GAO-15-291R.

[8] The SDR does not provide information on whether someone was funded by NASA or USDA, so we could not analyze these two agencies using the SDR. Additionally, the SDR does not provide information on whether someone applied for funding, only whether their work is supported by a grant or contract from a given agency. See appendix II for a detailed description of the SDR results and limitations associated with the data.

[9] See GAO-15-291R.

[10] Some agencies were only able to provide us data for 2-3 years because they did not collect it prior to a certain year. In addition, both NIH and NSF already calculate and publish success rates and other detailed data broken down by researcher sex, so these two agencies sent us summary data rather than individual records of proposals and awards.

[11] Ultimately, we chose not to present information on award amounts due to the lack of comparability in data from different agencies.

[12] In computing this statistic, revisions in a given year were not counted as separate applications.

[13] We requested both data on researcher sex and researcher first name from USDA for the purposes of conducting a robustness check. That is, we compared our gender name match to

the self-reported sex of researchers in the USDA dataset to see how well it performed against a dataset that included the researcher's sex.

[14] Other researchers have used this approach, most notably Susan D. Hosek, et al, *Gender Differences in Major Federal External Grant Programs* (Santa Monica: RAND Corporation, 2005).

[15] Our match did not perform as well for names which were less likely to be in the SSA database, or for gender neutral names as we required that a name be assigned exclusively to children of one sex or the other 95 percent of the time in order to be designated as a gender-specific name. Our analysis may not capture systematic differences in success rates by sex among individuals whose names do not appear in the database—for example, foreign-born researchers.

[16] The proportion of records for which we were able to assign a sex in the context of the total population of records could also be considered in the context of survey response and non-response measurement. In survey methodologies, the higher the response rate to a survey the less concerned one would be that the respondent population differs systematically from the non-respondent population. In this case, we were able to assign a sex to, at a minimum, more than two thirds of the population of records and approximately 75 percent of the population at some agencies—this means that less than one third of the records would be analogous to a "non-respondent" population if this were a survey. In these terms, this would be generally accepted as a reasonable response rate for this particular methodological approach.

[17] See GAO/AIMD-00-21.3.1.

[18] This objective focused on possible federal actions agencies could take beyond Title IX enforcement efforts to address the representation of women in federal STEM research.

[19] We narrowed our initial list from forty-six to twenty-five individuals who we judged to be familiar with the federal role based on their experience holding a national-level professional position, academic leadership position, or having co-authored a relevant study or report.

[20] This action did not surface during our initial literature review, but we became aware of it during the course of our interviews with federal agencies.

End Notes for Appendix II

[1] Respondents to the SDR may indicate that their work is funded by the federal government and, if so, they may select from a list one or more agencies from which they receive funding. The list is not an exhaustive list and includes options for NIH, NSF, DOD, and DOE, but not NASA or USDA.

[2] The 2013 SDR survey data are from a sample of 30,696 doctorate recipients. The tables and figures below apply weights to the data and reflect estimates, and estimated gender differences, in the underlying population of roughly 837,900 doctorate recipients in STEM fields from which the sample was drawn.

[3] Based on agency administrative data, success rates for women and men at NSF are approximately equal, suggesting the lower likelihood of funding for women at NSF may be due to factors which could not be controlled for in this study, such as lower propensity to apply for NSF funding among qualified female PhDs than among their male counterparts.

[4] Odds ratios of 1.0 indicate no difference. Thus odds ratios diminish as they approach 1.0, not as they approach 0.

In: Federal Grant-Making to Women … ISBN: 978-1-63485-679-9
Editor: Rochelle Schuler

Chapter 2

WOMEN IN STEM RESEARCH: FEDERAL AGENCIES DIFFER IN THE DATA THEY COLLECT ON GRANT APPLICANTS*

U.S. Covernment Accountability Office

Since the enactment of Title IX in 1972—which prohibits discrimination on the basis of sex in education programs and activities receiving any federal financial assistance[1]—women have made significant gains in many academic fields. Nevertheless, recent research shows that women continue to lag behind men in academic and professional advancement in science, technology, engineering, and math (STEM) fields. In fiscal year 2014, colleges and universities around the country received nearly $25 billion in federal funding for research in these fields.[2] You requested data regarding federal grant-making to women and men for research in STEM fields. This report provides the results of the first part of an ongoing review related to the representation of women in federal STEM research programs. Specifically, this report focuses on the extent to which federal agencies collect data that could be used to analyze differences, if any, in federal grants to women and men in STEM fields.[3]

We selected for review the six federal agencies that together funded approximately 90 percent of the federal government's investment in basic and

---

* This is an edited, reformatted and augmented version of The United States Government Accountability Office publication, GAO-15-291R, dated March 17, 2015.

applied research in fiscal year 2010 through 2012: the Department of Agriculture-National Institute of Food and Agriculture (USDA-NIFA), the Department of Defense (DOD), the Department of Energy (DOE), the National Aeronautics and Space Administration (NASA), the National Institutes of Health[4] (NIH), and the National Science Foundation (NSF).[5] At these six agencies, we collected data from 16 different entities that administer grant programs for research in STEM fields.[6] See figure 1 for the amount of research funding these six agencies provided to colleges and universities during fiscal year 2012, the most recent year for which complete agency-level data are available. We asked each agency to identify programs that fund research in STEM fields distributed through grants or cooperative agreements because these are the primary funding mechanisms by which colleges and universities receive research funds. A list of all the agencies and components included in this review can be found in enclosure I.

To conduct this work, we collected and reviewed relevant documentation from the agencies and components and interviewed agency officials. Specifically, we obtained and analyzed documentation about the data systems and data elements routinely tracked electronically on grant proposals and awards for research in STEM fields. We interviewed agency officials to understand how the data systems interface and how these data are maintained and used by the agencies for analysis. We assessed the reliability of the NSF data on the amount of research funding provided by federal agencies by reviewing documentation about how NSF's National Center for Science and Engineering Statistics collects and ensures the accuracy of these figures. We determined these data were sufficiently reliable for the purposes of this report.[7] See enclosure I for more details on our scope and methodology.

We conducted this performance audit from March 2014 to March 2015 in accordance with generally accepted government auditing standards. These standards require that we plan and perform the audit to obtain sufficient, appropriate evidence to provide a reasonable basis for our findings and conclusions based on our audit objectives. We believe the evidence obtained provides a reasonable basis for our findings and conclusions based on our audit objectives.

SUMMARY OF FINDINGS

Three of the six federal grant-making agencies we reviewed—the National Institutes of Health, the National Science Foundation, and the Department of

Agriculture-National Institute of Food and Agriculture—routinely collect demographic information about the individual researchers who apply for and receive federal grants. Further, these agencies have data systems that could facilitate analysis of differences, if any, in grants made to women and men. The three agencies stated they use this demographic information for research and internal analysis of their applicant pool. The other three agencies in our review—the Department of Defense, the Department of Energy, and NASA—do not routinely collect demographic information about researchers who submit grant proposals and receive awards. These agencies or some of their components reported they were uncertain of the legal constraints around collecting these data and that their agencies had no internal purpose for such information.

BACKGROUND

Federal Agencies That Fund Research in STEM Fields

The six federal agencies included in our analysis provide billions of dollars annually for research at colleges and universities in STEM fields. This report will focus on federal funding for research in the fields of engineering; life sciences (e.g., agricultural, biological, and environmental sciences); physical sciences (e.g., chemistry, earth sciences, and physics); computer and information technology; and math and statistics; where our prior work has shown women are still underrepresented.[8] Figure 1 details the total amount of research funding provided in these STEM fields by each agency in fiscal year 2012.

Data to Calculate Success Rates and Analyze Differences between Women and Men in Grant Awards for Research in STEM Fields

The success rate is one statistic that allows for a comparison of federal grant-making between different groups of applicants, such as women and men.[9] The success rate is the number of grants awarded divided by the number of grant proposals received.[10]

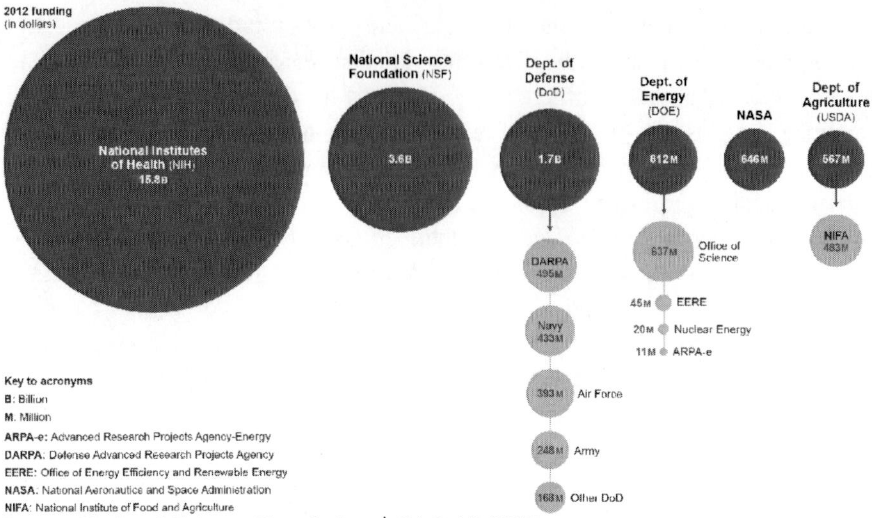

Source: National Science Foundation. | GAO-15-291R.

Notes:The funding totals presented in figure 1 include only research obligations in fields that fall within the scope of our review (i.e., computer and information technology, engineering, environmental sciences, life sciences, math and statistics, and physical sciences), and may differ from total agency research funding obligations for FY 2012. For example, NSF's total STEM research obligations include funding for additional STEM fields such as psychology and social sciences. FY 2012 is the most recent year for which complete data are available for the agencies and components included in this review.

Figure 1. Fiscal Year 2012 Federal Obligations for Select Areas of STEM Research Performed at Universities and Colleges.

This calculation is important because simply comparing the total number of grants awarded to women to the total number of grants awarded to men does not take into account that there are fewer women participating in STEM fields.[11] In order to calculate success rates for women and men, data on the sex of the grant applicant—or principal investigator (PI)—needs to be collected and retained for both proposals and awards.[12] Ideally, each grant proposal would be assigned a unique ID that would also be assigned to any award that resulted from the proposal. This linkage between the record of the grant proposal and award ensures that for every grant award in the numerator of the success rate, there is a corresponding grant proposal in the denominator of the success rate. A data system that integrates or electronically links proposal and award data facilitates this type of analysis.

Calculating the success rate alone, however, does not produce complete and accurate information about what differences may exist between women and men receiving federal funding for research in STEM fields. First, it is also important to compare the size of grants awarded to women and men. Second, there may be other characteristics of the female and male applicants for research grants that influence the likelihood of success or award size. To determine if sex alone has any effect on whether an applicant is awarded a grant or on the size of the grant, one needs to compare women and men with similar educational and career characteristics. Therefore, federal agencies would need to collect data on these characteristics in order to accurately determine whether an individual's sex seems to have a significant effect on the likelihood of receiving a grant.[13] Including variables on these types of characteristics in any analysis could be used to explain the source of differences, if any, between funding to women and men and could also clarify what proportion of any differences remains unexplained after controlling for educational and career characteristics.

Previous studies that have analyzed the factors affecting federal grant funding have also made use of data on researcher characteristics. Specifically, such studies have used demographic characteristics of principal investigators such as age, race, and sex; educational characteristics, such as highest degree and discipline of degree; and career characteristics, such as faculty position, years of experience, tenure status, and type and/or rank of the institution from which the investigator is applying.[14] For example, in 2005, the RAND Corporation completed a study on behalf of NSF that analyzed sex differences in federal grant-making using such variables to explain differences in grants awarded to women and men at NIH, NSF, and USDA.[15] In their report, RAND cited the lack of adequate data collection and other limitations in the data collected by federal agencies that fund research in STEM fields as a major impediment to conducting their analyses.[16] See enclosure I for a list of key variables that we identified as important to studying potential factors affecting grant funding differences between women and men from our review of previous studies.

THREE OF SIX AGENCIES ROUTINELY COLLECT DATA THAT ALLOW FOR ANALYSIS OF POTENTIAL DIFFERENCES IN FEDERAL GRANT-MAKING BETWEEN WOMEN AND MEN

Three of Six Agencies Collect Data on the Demographic Characteristics of Applicants and Awardees

Three of the six agencies we reviewed routinely collect information on researcher characteristics that could be used to analyze differences in federal grants and cooperative agreements between women and men for research in STEM fields.[17] While some components in all six agencies collect some data on researcher characteristics that would be helpful in conducting this type of analysis, only NIH, NSF, and USDA-NIFA routinely collect data on the sex of applicants and grantees, as well as a variety of other demographic and career characteristic variables that could be used to explain the gaps, if any, between women and men.[18] While these three agencies do not require applicants to identify themselves as male or female, the percentages of individuals who voluntarily provide this information in response to demographic questions in the application are typically high.[19] Both NIH and NSF maintain separate researcher records in their database with a unique researcher identifier to allow the individual's profile to be associated with multiple proposals and awards over time. Demographic data are not used as a part of the proposal review process; rather, their purpose is primarily to allow the agencies to analyze their own applicant and grantee pool. For example, NSF recently published a study of its own merit review process that analyzes the applicant and grantee pool by demographic characteristics.[20] In addition, these data have been used for other research Institution purposes. For example, NIH data have been used in previous studies on both the effect of a primary investigator's sex and race on the likelihood of receiving a grant, and NIH regularly reports on the demographic breakdown of its grant applicants and recipients.[21] Although USDA was included in the 2005 RAND study of this topic, USDA-NIFA has not recently published its own statistics on the characteristics of its applicants and grantees.

In contrast, the other three agencies—DOE, DOD, and NASA—do not collect data on whether grant applicants are male or female and vary in the extent to which they collect other researcher characteristics. Specifically, one of four components at DOE and three of eight components at DOD included in our review do not retain any electronic data on applicants, making calculation of a grant success rate for these components impossible. On the other hand, NASA and several components at DOE and DOD do collect data on some researcher characteristics (e.g., degree and discipline).[22] Table 1 lists some of the relevant data agencies collect on researcher characteristics.

Table 1. Researcher Information Collected by Agency Administrative Data Systems

Variables collected	U.S. Dept. of Agriculture[a]	Dept. of Defense[b]	Dept. of Energy[c]	NASA	National Institutes of Health	National Science Foundation
Sex	✓				✓	✓
If tracked, approx. % of records with sex recorded[d]	86%				90-95%	85-90%
Race	✓				✓	✓
Age/year of degree	✓	1 of 8 components			✓	✓
Institution	✓	✓	✓	✓	✓	✓
Highest degree	✓	2 of 8 components			✓	✓
Discipline (or subject area of grant)	✓	4 of 8 components	3 of 4 components	✓	✓	
Co-Investigator information[e]	✓	2 of 8 components		✓	✓	

Source: GAO analysis of agency documents. | GAO-15-291R

Notes: We did not evaluate the extent to which these data fields were populated (other than for the data on applicants' sex, where collected), nor did we evaluate the quality or reliability of the data the agencies collected.

[a]The only component included in our review at USDA was the National Institute of Food and Agriculture (NIFA).

[b]Three of eight components included in this review at DOD do not retain proposal data for grant applicants once awards are made.

[c] One of four components included in this review at DOE does not retain proposal data for grant applicants once awards are made.

[d] Demographic information such as sex and race are collected on a voluntary basis only; applicants are not required to provide this information to the agency. The percentages in this table were reported to us by the agencies. NIH and NSF provided us a range based on the past several fiscal years and USDA-NIFA provided us the percentage from their most recent full year of data, FY 2013.

[e] In this row, agencies received a checkmark if they collected any information regarding co-investigators, such as name or institution.

The three agencies we reviewed that routinely collect demographic data—NIH, NSF, and USDA-NIFA—each have an integrated data system that enables the agency to track the complete lifecycle of a single grant, and to link a single proposal with the award decision. Having linked proposal and award data enables the agency to compute and analyze success rates. Consequently, these agencies have data systems that could facilitate analysis of differences, if any, in grants made to women and men.

The remaining three agencies that do not routinely collect demographic data across all of their components—DOD, DOE, and NASA—have data systems that vary in the extent to which they integrate proposal and award data. In some cases, these data systems only partially or minimally integrate proposal and award data.[23] This makes it difficult to reliably link a single proposal with the award decision, and therefore to calculate a reliable grant success rate. For example, NASA and some components at DOD and DOE use separate databases that are not electronically linked, some of which retain data on proposals and others for grants awarded. In addition, some components at DOD and DOE use minimally integrated systems that retain no electronic data at all on applicants and track only minimal data on awardees. At DOE for example, the Office of Nuclear Energy does not retain data on grant applicants, only on awardees. At DOD, the Basic Research Office within Acquisition, Technology & Logistics (AT&L) also does not retain applicant data.[24] Without information on the number of applicants and information about those applicants, one cannot calculate a success rate overall or by gender or award type. See figure 2 for types of systems agencies use to collect data.

Agencies Cited a Variety of Reasons for Why They Do or Do Not Collect Demographic Data

Agencies reported several reasons for why they do or do not routinely collect these types of data. For example, four agencies and seven components told us they did not collect this information because there was no legal requirement to collect demographic data on applicants or grantees.[25] Further, one agency and five components cited concerns about whether they are allowed to collect demographic data. Officials at one agency stated that they had been advised by their agency's general counsel not to ask demographic questions on their grant applications.

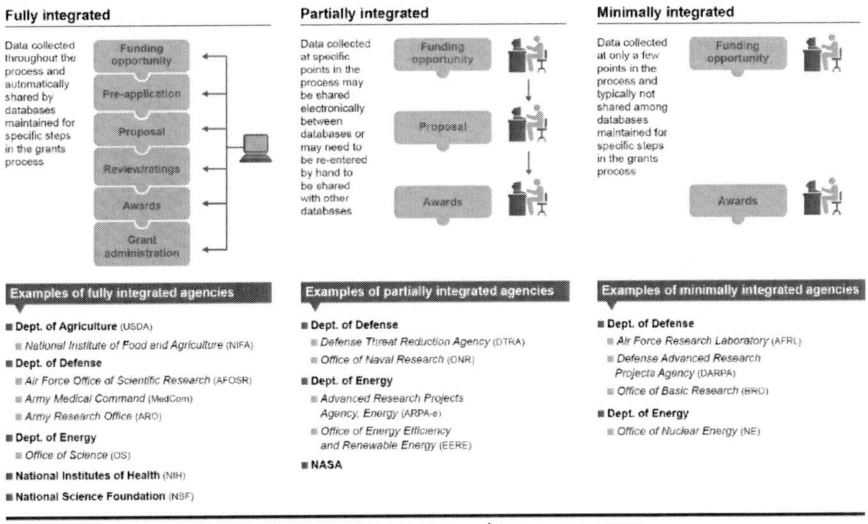

Source: GAO analysis of data from the agencies. | GAO-15-291R.

Note: Relevant components are only listed in this graphic where data systems differ within the agency.

Figure 2. Types of Agency Data Systems Used to Collect Data on Grant Proposals and Awards.

In contrast, the three agencies that have been collecting these data for a number of years—NIH, NSF, and USDA-NIFA—cited the desire to monitor their award processes as the basis for collecting these data. Both USDA-NIFA and NSF use the same personal data collection form approved by the Office of Management and Budget to collect demographic data, which provides the following guidance to applicants:

> "The Federal Government has a continuing commitment to monitor the operation of its review and award processes to identify and address any inequities based on gender, race, ethnicity, or disability of its proposed PDs/PIs [Project Directors/ Principal Investigators] and co-PDs/PIs . . . . Submission of the requested information is voluntary and is not a precondition of award . . . . Upon receipt of the application, this form will be separated from the application. This form will not be duplicated, and it will not be a part of the review process."

Similarly, NIH offers the following guidance to applicants concerning their voluntary submission of demographic data:

> "Federal Agencies have a continuing commitment to monitor the operation of its [sic] review and award processes to detect, and deal appropriately with, any instances of real or apparent inequities . . . . Personal demographic data on PD/PIs [Project Directors/Principal Investigators] and those with a postdoctoral role is vital to comply with these requirements."

Examples of data collection forms from these three agencies are included in enclosure II.

Agencies also varied in their capacity and willingness to collect data they deem peripheral to the grant decision making process. Two agencies and three components from a third agency indicated that they lacked the capacity in existing systems to collect these data. For example, officials at one component stated that the standard forms available from Grants.gov—a centralized electronic portal for applicants to find information about government funding opportunities—provided no mechanism for asking applicants for demographic information. However, other agencies include demographic data forms or links to such forms on Grants.gov.[26] Lastly, when we asked agencies and components why they did not collect certain data elements, the most common response from agencies and components was that they had no internal use for such data.

AGENCY COMMENTS

We provided a draft of this report to the departments of Agriculture, Defense, Energy, Health and Human Services (specifically, the National Institutes of Health), the National Aeronautics and Space Administration, and the National Science Foundation. DOE, NASA, and NSF provided technical comments, which we incorporated as appropriate. USDA, DOD, and NIH had no comments on the draft.

Melissa Emrey-Arras
Director,
Education, Workforce, and Income Security

Enclosures – 2

ENCLOSURE I: SCOPE AND METHODOLOGY

We reviewed the total amount of federal research and development funding from fiscal year 2010 through 2012 and selected the top six federal funding agencies to include in our review: the Department of Agriculture-National Institute of Food and Agriculture (USDA-NIFA), the Department of Defense (DOD), the Department of Energy (DOE), the National Aeronautics and Space Administration (NASA), the National Institutes of Health (NIH), and the National Science Foundation (NSF).[27] These six agencies funded more than 90 percent of basic and applied research in STEM fields in 2012. We assessed the reliability of the NSF survey data on federal funds for research and development by reviewing documentation about how the National Center for Science and Engineering Statistics collects these data from federal agencies and ensures the accuracy of these figures. We determined these data were sufficiently reliable for the purposes of our review. See table 2 for a list of all of the federal agencies and components we included.

Table 2. Agencies and Components in Our Review

Agency	Agency Component
Department of Agriculture (USDA)	National Institute of Food and Agriculture
Department of Defense (DOD)	Air Force Office of Scientific Research Air Force Research Laboratory Army Medical Command Army Research Office Basic Research – Office of Assistant Secretary of Defense for Research and Engineering Defense Advanced Research Projects Agency Defense Threat Reduction Agency Office of Naval Research
Department of Energy (DOE)	Advanced Projects Agency (Energy) Office of Energy Efficiency and Renewable Energy Nuclear Energy Office of Science
National Aeronautics and Space Administration (NASA)	
National Institutes of Health (NIH)	
National Science Foundation (NSF)	

Source: GAO. | GAO-15-291R

At each of these agencies and components, we provided a data collection instrument that asked officials to identify all programs that met the following inclusion criteria for our review:

1. Research funding in core STEM fields only: computer and information technology, engineering, math, physical and life sciences, and statistics.
2. Funding for basic and applied research at the graduate, post-graduate, and principal investigator or full professor level.
3. Funds distributed through external grants or cooperative agreements.

We limited our scope to external grants or cooperative agreements because these are the primary funding methods by which colleges and universities receive research funds. Officials at each agency and component we included identified the relevant STEM research programs at the agency level and, across these six agencies, we identified 16 different entities that administered these research grant-making programs with varying degrees of independence, depending on the agency's structure.

We conducted an extensive literature review to identify previous studies that have analyzed factors affecting whether a researcher receives federal funding. Such studies typically include in their analysis demographic characteristics such as age, race, and sex; educational characteristics such as highest degree, discipline of degree; and career characteristics such as tenure status, rank of current institution, and years of experience. Based on these studies and our own analysis, we developed a list of variables that would be useful in calculating success rates for women and men and to analyze what factors, including the sex of the principal investigator, could possibly affect whether a grant proposal is selected for funding at a federal agency. See table 3 for a list of key variables we identified through our own analysis and a review of the relevant literature.[28]

Table 3. Key Variables Useful in Analyzing Success Rates and Explaining Any Differences in Success Rates between Women and Men

Variables Used in Previous Studies	Purpose
Age	Experience proxy
Amount awarded	Main dependent variable
Amount requested	Main dependent variable
Award status (approved, declined)	Main dependent variable

Variables Used in Previous Studies	Purpose
Award status (approved, declined)	Main dependent variable
Begin date of award	To calculate fiscal year
Degree	Explanatory variable
End date of award	To calculate fiscal year
Experience	Explanatory variable
Fiscal year of award	Explanatory variable
Institution	To identify institutional rank
Institutional rank	Explanatory variable
Institution type (college vs. research university)	Explanatory variable
Nonacademic applicant	Explanatory variable
Not PhD	Explanatory variable
Proposal receipt date	To calculate fiscal year
Discipline	Explanatory variable
Sex	Focus of research question: to analyze if the sex of applicant is a significant determinant
Tenure track/status	Explanatory variable
Type of proposal (reapply, modification, initial)	To filter out out-of-scope awards
Identified by Previous Studies as a Meaningful Explanatory Variable	
Award type	Explanatory variable, to filter/analyze awards by type
Co-Investigator data	Explanatory variable
Ethnicity	Explanatory variable
Race	Explanatory variable
Reviewer score	Explanatory variable
Type of academic appointment	Explanatory variable
Year of degree	Explanatory variable
Further Data Useful For Summary Grant Statistics	
Amount awarded for each corresponding award, gender	Total and mean amount awarded, total and by gender
Amount requested in each proposal, gender	Total and mean amount requested, total and by gender
Award unique ID, link to unique proposal ID, sex	To calculate success rates
Award type, unique award ID	All of the above by award type
Proposal unique ID	To calculate success rates

Source: GAO analysis and review of literature. | GAO-15-291R.

In order to determine whether agencies were collecting information on these characteristics, we interviewed officials at the agency level and at each of the relevant components to understand their grants management and procurement databases. We verified this information by reviewing documents from each agency and component that explained these data systems and what data their systems collected. We reviewed documents such as:

- Data dictionaries
- Database handbooks
- Database table schema
- Forms used to collect data

Reviewing these documents, we identified the relevant data fields each agency or component collected and compared this information with the list of key variables we identified in our analysis of previous studies. Based on data fields we observed to be absent from data dictionaries and other documents, we created lists of relevant data each agency and component did not collect. We then confirmed the list of data not collected by a particular agency or component with officials at that particular agency or component. We also asked officials from each agency and component to provide the main reasons why they do or do not collect these data. It is important to note that this report focused on whether the funding agencies have existing data systems and procedures in place to collect these data and we did not evaluate the extent to which these data fields were populated, nor did we evaluate the quality or reliability of the data the agencies collected.

ENCLOSURE II: EXAMPLES OF FORMS AGENCIES USE TO COLLECT DEMOGRAPHIC DATA

Three agencies collect sex and other demographic characteristics of grant applicants. Both the National Science Foundation and the Department of Agriculture-National Institute of Food and Agriculture use a standard form approved by the Office of Management and Budget (OMB) for collecting these data as part of their proposal process. See figure 3.

OMB Number: 4040-0001
Expiration Date: 6/30/2016

RESEARCH & RELATED PERSONAL DATA

Project Director/Principal Investigator and Co-Project Director(s)/Co-Principal Investigator(s)

The Federal Government has a continuing commitment to monitor the operation of its review and award processes to identify and address any inequities based on gender, race, ethnicity, or disability of its proposed PDs/PIs and co-PDs/PIs. To gather information needed for this important task, the applicant should submit the requested information for each identified PD/PI and co-PDs/PIs with each proposal. Submission of the requested information is voluntary and is not a precondition of award. However, information not submitted will seriously undermine the statistical validity, and therefore the usefulness, of information received from others. Any individual not wishing to submit some or all the information should check the box provided for this purpose. Upon receipt of the application, this form will be separated from the application. This form will not be duplicated, and it will not be a part of the review process. Data will be confidential.

Project Director/Principal Investigator

Prefix: *** First Name:** **Middle Name:**

*** Last Name:** **Suffix:**

Gender:

Race (check all that apply):
- American Indian or Alaska Native
- Asian
- Black or African American
- Native Hawaiian or Other Pacific Islander
- White
- Do Not Wish to Provide

Ethnicity:

Disability Status (check all that apply):
- Hearing
- Visual
- Mobility/Orthopedic Impairment
- Other
- None
- Do Not Wish to Provide

Citizenship:

Next Person

Source: GUSDA-NIFA Application Form. |GAO-15-291R.

Figure 3. OMB Form Used by the National Science Foundation and the United States Department of Agriculture-National Institute of Food and Agriculture for Collecting Applicant Demographic Data.

The National Institutes of Health collects personal data through the Electronic Research Administration (eRA) Commons, its electronic data collection and grants administration tool. Specifically, data are collected through the Personal Profile. The data are provided by the individual through a secure, electronic system; an example form is shown in Figure 4.

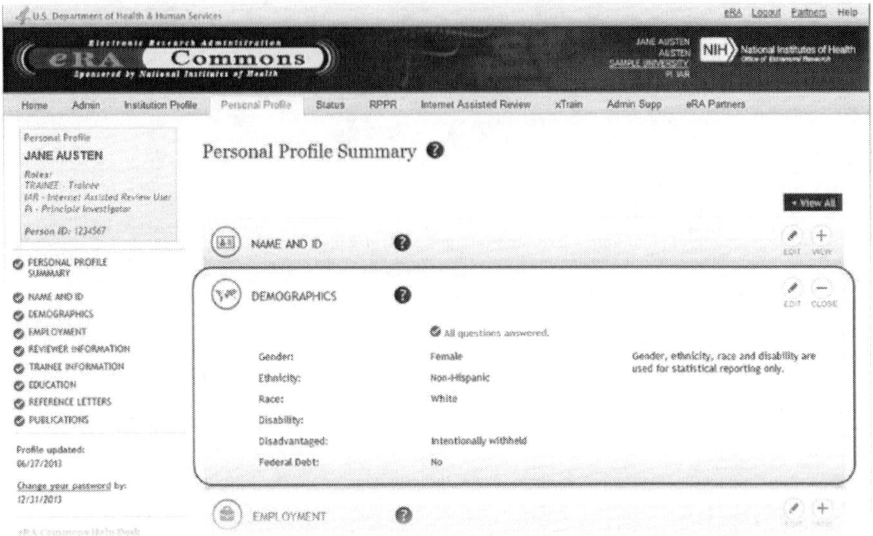

Source: NIH eRA Commons Personal Profile Module User Guide. | GAO-15-291R.

Figure 4. National Institutes of Health Form for Collecting Applicant Demographic Data.

End Notes

[1] Pub. L. No. 92-318, tit. IX, § 901, 86 Stat. 235, 373.

[2] This figure is reported annually by the National Science Foundation's National Center for Science and Engineering Statistics. The fiscal year 2014 figure is preliminary and includes the total amount of federal obligations for research at universities and colleges in the fields of computer science and mathematics, engineering, environmental sciences, life sciences, and physical sciences.

[3] Federal research grants are typically awarded to the institution (e.g., college or university) that employs the researcher and not directly to the individual researcher. In this report, we primarily focused on data that federal agencies collect about characteristics of the principal investigators that submit proposals for grant funding, not the institution that employs them.

[4] According to 2012 figures, NIH, an agency of the Department of Health and Human Services, was responsible for administering more than 90 percent of the department's re search budget. Therefore, we focused on NIH in this report.

[5] Basic research concerns fundamental aspects of phenomena without specific applications toward products or processes in mind; applied research concerns gaining understanding necessary to determine the means by which a specific need may be met.

[6] The entities we reviewed within each agency vary and may be referred to as offices, commands, or components. For the purposes of this report, we will refer to these separate entities within a larger agency as "components".

[7] This report focused on whether the funding agencies have existing data systems and procedures in place to collect data. We did not evaluate the extent to which these data fields were populated, nor did we evaluate the quality or reliability of the data the agencies collected.

[8] GAO, Science, Technology, Engineering, and Mathematics Education: Assessing the Relationship between Education and the Workforce, GAO-14-374 (Washington, D.C.: May 8, 2014).

[9] Total number of awards, total dollar amounts of awards, and average dollar amount of awards are also relevant statistics to compare funding differences between women and men.

[10] To calculate the number of grant proposals used to calculate the success rate, revisions to a grant are counted as part of the same grant application. The "award rate" is the term used if the number of grants awarded is divided by number of applications received where revisions are counted as separate applications.

[11] According to NSF data, in fiscal year 2012, women made up around 35 percent of the graduate student and postdoctoral fellow population in STEM fields, and women were awarded 38.8 percent of doctoral degrees in STEM fields.

[12] Depending on the agency, grants can have multiple principal investigators, or co-investigators may also be designated. Previous research has suggested that ideally, the sex of all principal investigators and co-investigators would be recorded for every grant proposal received. For example, see Susan D. Hosek, et al, Gender Differences in Major Federal External Grant Programs (Santa Monica: RAND Corporation, 2005).

[13] The same information would be required to determine whether other researcher characteristics (e.g., age, disability, race, etc.) have an effect on the likelihood of receiving a grant.

[14] Academic institutions can be categorized using different classification systems. For example, the Carnegie classification system categorizes all colleges and universities in the U.S. based on types of degrees granted, population served and other characteristics. Other systems classify institutions by research rankings, such as those developed by The Center for Measuring University Performance, which divides research universities into different tiers based on research output and other characteristics.

[15] The National Science Foundation Authorization Act of 2002 (Pub. L. No. 107-368, § 18(c), 116 Stat. 3034, 3061), directed NSF to sponsor a study to "examine differences in amounts requested and awarded, by gender, in major Federal external grant programs". See Susan D. Hosek, et al, Gender Differences in Major Federal External Grant Programs (Santa Monica: RAND Corporation, 2005).

[16] Both the RAND report and the National Academies of Science report Beyond Bias and Barriers: Fulfilling the Potential of Women in Academic Science and Engineering (Washington, D.C.: The National Academies Press, 2006) identified lack of data as an impediment to research in this area. Other studies of federal grant-making in STEM fields focused exclusively on NIH, which does collect data on researcher characteristics. See Pohlhaus, Jennifer Reineke, et al, "Sex Differences in Application, Success, and Funding Rates for NIH Extramural Programs." Academic Medicine, vol. 86, no. 6 (2011). See also Ginther, Donna K., et al, "Race, Ethnicity, and NIH Research Awards", Science, vol. 333, no. 1015 (2011).

[17] These three agencies accounted for approximately 85 percent of the $25 billion in total federal STEM research funding to colleges and universities.

[18] We developed a list of variables needed to analyze factors that influence grant success by reviewing the methodology of relevant studies. For example, demographic variables include age, race, and sex, and career characteristic variables include discipline or highest degree attained. A list of key variables identified in these studies is included in enclosure I.

[19] According to agency officials at USDA-NIFA, in FY 2013, approximately 86 percent of grant records have the gender field populated; officials at NIH and NSF said the percentage is between 90-95 percent and 85-90 percent, respectively, over the past several years.

[20] National Science Foundation, Report to the National Science Board on the National Science Foundation's Merit Review Process Fiscal Year 2013, NSB-14-32 (May 12, 2014).

[21] Ginther, Donna K., et al, "Race, Ethnicity and NIH Research Awards", Science, vol. 333, no. 1015 (2011). Pohlhaus, Jennifer Reineke, et al, "Sex Differences in Application, Success, and Funding Rates for NIH Extramural Programs", Academic Medicine, vol. 86, no. 6, (June 2011). Ley, Timothy J. and Barton H. Hamilton, "The Gender Gap in NIH Grant Applications", Science, vol. 332 (2008). See also National Institutes of Health, NIH Data Book, http://report.nih.gov/nihdatabook/.

[22] NASA, through its Office of the Chief Scientist, is taking steps to pursue the feasibility of collecting additional demographic data (including age, race, and sex) about applicants to its competitive grant solicitations. NASA officials anticipate rolling out the implementation of this data collection effort in the fall of 2015.

[23] Some components at DOD have integrated data systems. For example, though it does not routinely collect demographic information about its grant applicants, in 2014, the Air Force Office of Scientific Research at DOD began tracking proposals, awards, and grants in one system.

[24] Officials at DOD told us that the Basic Research Office (in AT&L) makes approximately 10 awards per year. Minimal data on successful grantees is tracked in a spreadsheet; however, because of the small number of awards, officials said it would be feasible for the program manager in the Basic Research Office to provide detailed data on the grantees without the need for a formal database.

[25] Title IX, the federal statutory provision prohibiting discrimination on the basis of sex in education programs and activities receiving any federal financial assistance, does not require federal agencies to collect any data on the sex of grant applicants or grantees. According to officials at the Department of Education's Office for Civil Rights and the Department of Justice's Civil Rights Division, while Title IX does not require federal agencies to collect this type of data, it also does not prohibit it.

[26] An E-Government initiative under the governance of the Office of Management and Budget, Grants.gov contains a repository of standard data collection forms for agencies' use in building proposal packages for funding opportunities, and agencies are able to customize these application packages, which can interface with their existing grant management systems. In addition, federal agencies can post funding opportunities and receive proposals through Grants.gov.

[27] According to 2012 figures, NIH, an agency of the Department of Health and Human Services, was responsible for administering more than 90 percent of the department's research budget. Therefore, we focused on NIH in this report.

[28] See, for example, Susan D. Hosek, et al, Gender Differences in Major Federal External Grant Programs (Santa Monica: RAND Corporation, 2005) and Ginther, Donna K., et al, "Race, Ethnicity, and NIH Research Awards", Science, vol. 333, no. 1015 (2011).

INDEX

A

B

C

D

E

F

G

H

I

J

L

M

N

O

P

U

V

W